PREPARING FOR THE BIOLOGY
AP*EXAM

WITH
BIOLOGY
CAMPBELL/REECE

TEXT PLUS TEST

PEARSON
SERIES
FOR
AP*
SUCCESS

*AP is a registered trademark of the College Board, which was not involved in the
production of, and does not endorse, this product.

PEARSON

Prentice
Hall

Figures taken from:

Biology, Sixth Edition
By Neil A. Campbell and Jane B. Reece
Copyright © 2002 by Pearson Education, Inc.
Published by Benjamin Cummings
San Francisco, California 94111

Test Bank for Biology, Sixth Edition
edited by William Barstow
Copyright © 2002 by Pearson Education, Inc.
Published by Benjamin Cummings

Student Study Guide for Biology, Sixth Edition
by Martha R. Taylor
Copyright © 2002 by Pearson Education, Inc
Published by Benjamin Cummings.

This special edition published in cooperation with Pearson Custom Publishing.

Printed in the United States of America

10 9 8 7 6 5 4 3 2

ISBN 0-536-73156-X

BA 998575

JC/NN

Please visit our web site at *www.pearsoned.com*

PEARSON PRENTICE HALL
Upper Saddle River, New Jersey 07458
A Pearson Eductaion Company

Preparing for the AP* Biology Examination with Campbell/Reece: *Biology,* Sixth Edition

TABLE OF CONTENTS

About Your
Pearson Text Plus Test AP* Guide

Pearson Education is the leading publisher of textbooks worldwide. With operations on every continent, we make it our business to understand the changing needs of students at every level, from Kindergarten to college. We think that makes us especially qualified to offer this series of AP test prep books, tied to some of our best-selling textbooks.

Our reasoning is that as you study for your course, you're preparing along the way for the AP test. If you can tie the material in the book directly to the test you're taking, it makes the material that much more relevant, and enables you to focus your time most efficiently. And that's a good thing!

The AP exam is an important milestone in your education. A high score means you're in a better position for college acceptance, and possibly puts you a step ahead with college credits. Our goal is to provide you with the tools you need to excel on the exam . . . the rest is up to you.

Good luck!

Part I

Introduction to the AP Biology Examination

This section gives an overview of the Advanced Placement* program and the AP Biology Examination. Part I introduces the types of questions you will encounter on the exam, provides helpful test-taking strategies, and explains the procedures used to grade the exam. Finally, a correlation chart shows where in Campbell and Reece, *Biology,* Sixth Edition, you will find key information that commonly appears on the AP Biology Examination. Review Part I carefully before trying the sample test items in Part II and Part III.

The Advanced Placement Program*

Probably you are reading this book for a couple of reasons. You may be a student in an Advanced Placement (AP) Biology class, and you have some questions about how the whole AP Program works and how it can benefit you. Also, perhaps, you will be taking an AP Biology Examination, and you want to find out more about *that*. This book will help you in several important ways. The first part of this book introduces you to the AP Biology course and the AP Biology Exam. You'll learn helpful details about the different formats—multiple-choice questions and free-response questions—that you'll encounter on the exam. In addition, you'll find dozens of test-taking strategies that will help you prepare for the exam. A correlation chart at the end of Part I shows how to use your textbook, Campbell and Reece *Biology*, to find the information you'll need to know to score well on the AP Biology Exam. By the way, this chart is useful, too, in helping you to identify any extraneous material that *won't* be tested. Also in Part I, you'll find information that clarifies the criteria that will be used to evaluate your work on the exam. Part II of this book provides an extensive content review, along with sample multiple-choice and free-response questions. Finally, in Part III, you will find two full-length sample tests. These will help you practice taking the exam under real-life testing conditions. The more familiar you are with the AP Biology Exam ahead of time, the more comfortable you'll be on testing day. And the more comfortable you are, the better your chances of achieving a high score.

The AP Program is sponsored by the College Board, a nonprofit organization that oversees college admissions examinations. (The College Board is composed of college and high school teachers and administrators.) The AP Program offers thirty-five college-level courses to qualified high school students. If you receive a grade of 3 or higher on an AP exam, you may be eligible for college credit, depending on the policies of the institution you plan to attend. Approximately 3,000 college and universities around the world grant credit to students who have performed well on AP exams. If you are taking several AP courses and if you score well on multiple AP exams, you may even be eligible to enter college as a sophomore. Some institutions grant sophomore status to incoming first-year students who have demonstrated mastery of many AP subjects. In addition, the College Board confers a number of AP Scholar Awards on students who score 3 or higher on three or more AP exams. Additional awards are available to students who receive very high grades on four or five AP exams.

Why Take an AP Course?

You may be taking an AP course simply because you like challenging yourself and you are thirsty for knowledge. Another reason may be that you know that colleges look favorably on applicants who have AP courses on their secondary school transcripts. AP classes involve rigorous, detailed lessons, a lot of homework, and numerous tests. Your willingness to take them tells college admissions officers that you believe in working hard to get the most from your education. Because AP course work is more difficult than average high school work, many admissions officers evaluate AP grades on a higher academic level. For example, if you receive a *B* in an AP class, it might carry the same weight as an *A* in a regular high school class.

Your AP Biology course prepares you for many of the skills you will need in college. For example, your teacher may assign a major research paper and require you to perform several challenging laboratory exercises using proper scientific protocol. AP Biology teachers routinely give substantial reading assignments, and students learn how to take detailed lecture notes and participate vigorously in class discussions. The AP Biology course will challenge you to gather and consider information in new—and sometimes unfamiliar—ways. You can feel good knowing that your ability to use these methods and skills will give you a leg up as you enter college.

Each college or university decides whether or not to grant college credit for an AP course, and each bases this decision on what it considers satisfactory grades on AP exams. Depending on what college you attend and what area of study you pursue, your decision to take the AP Biology Exam could end up saving you lots of tuition money. You can contact schools directly to find out their guidelines for accepting AP credits.

Taking an AP Examination

The AP Biology Exam is given annually in May. Your AP teacher or school guidance counselor can give you information on how to register for an AP exam. Remember, the deadline for registration and payment of exam fees is usually in January, four months before the actual exam date in May. The cost of the exam is subject to change and can differ depending on the number of exams taken. However, in 2003 the cost of a single exam was $80. For students who can show financial need, the College Board will reduce the price by $22, and your school might also waive its regular rebate of $8, so the lowest possible total price is $50. Moreover, schools in some states are willing to pay the exam fee for the student. If you feel you may qualify for reduced rates, ask your school administrators for more information.

The exams are scored in June. In mid-July (about 6–8 weeks after you take the exam) the results will be sent to you, your high school, and any colleges or universities you've marked down on your answer sheet. If you want to know your score as early as possible, you can get them (for an additional charge of $15) beginning July 1st by calling the College Board at (609)-771-7300. On the phone, you'll be asked to give your AP number or social security number, your birth date, and a credit card number.

After you receive your score, if you decide that you want it sent to additional colleges and universities, you can fill out the appropriate information on your AP Grade Report (which you will receive by mail in July) and return it to the College Board. There is an additional charge of $14 for each additional school that will receive your AP score.

On the other hand, if a feeling of disaster prevents you from sleeping on the nights following the exam and you'd prefer to pretend the whole thing never happened, you can choose to withhold or cancel your grade. (Withholding is temporary, whereas canceling is permanent.) Each procedure carries a $5 charge per college or university. You'll need to write to (or email) the College Board and include your name, address, gender, birth date, AP number, the date of the exam, the name of the exam, a check for the exact amount due, and the name, city, and state of the college(s) from which you want the score withheld. You should check the College Board website for the deadline for withholding your score, but it's usually in mid-June. A word of advice: we strongly suggest that you *do not* cancel your scores, since you won't know your score until mid-June. We recommend sitting back, relaxing, and assuming that the glass is half full. At this point, you have nothing to lose and a lot to gain.

Two final notes. If you would like to get back your free-response booklet for a post-exam review, you can send *another* check for $10 to the College Board. You'll need to do this by mid-September. Finally, if you have serious doubts about the accuracy of your score for the multiple-choice section, the College Board will re-score it for an additional $15.

AP Biology: Course Goals

The two central goals of the AP Program in Biology are to help students develop a conceptual framework for modern biology and gain an appreciation of science as a process. AP Biology courses are built around topics, concepts, and themes. The College Board defines *topics* as the subject areas of biology. A *concept* is an important idea or principle that forms or enhances our current understanding of a particular topic. *Themes* are the overarching features of biology that recur, connect, and unify our understanding of topics. The College Board lists eight themes that should be stressed in AP Biology courses. The following outline is taken directly from "Topic Outline," page 5, AP Biology Course Description, published by the College Board:

▌ Science as a process;
▌ Evolution;
▌ Energy transfer;
▌ Continuity and change;
▌ Relationship of structure to function;
▌ Regulation;
▌ Interdependence in nature;
▌ Science, technology, and society.

During the year, AP Biology students will be applying those themes to a wide range of topics. In fact, the College Board has created a topic outline to illustrate the topics that make up a typical college biology course—and so should form the basis for AP Biology courses. The percentages in parentheses show how much of the course is spent on particular topics.

    I. Molecules and Cells (25% of the AP Biology course)
       A. Chemistry of Life (7%)
          Water
          Organic molecules in organisms

Free energy changes
Enzymes
B. Cells (10%)
Prokaryotic and eukaryotic cells
Membranes
Subcellular organization
Cell cycle and its regulation
C. Cellular Energetics (8%)
Coupled reactions
Fermentation and cellular respiration
Photosynthesis
II. Heredity and Evolution (25% of the AP Biology course)
A. Heredity (8%)
Meiosis and gametogenesis
Eukaryotic chromosomes
Inheritance patterns
B. Molecular Genetics (9%)
RNA and DNA structure and function
Gene regulation
Mutation
Viral structure and replication
Nucleic acid technology and applications
C. Evolutionary Biology (8%)
Early evolution of life
Evidence for evolution
Mechanisms of evolution
III. Organisms and Populations (50% of the AP Biology course)
A. Diversity of Organisms (8%)
Evolutionary patterns
Survey of the diversity of life
Phylogenetic classification
Evolutionary relationships
B. Structure and Function of Plants and Animals (32%)
Reproduction, growth, and development
Structural, physiological, and behavioral adaptations
Response to the environment
C. Ecology (10%)
Population dynamics
Communities and ecosystems
Global issues

No doubt, AP Biology courses vary somewhat from teacher to teacher and from school to school. For the most part, many high school teachers study the AP Biology course outline each year and meticulously customize their curriculum to fit it. In short, you may wish to consult the AP Biology course outline in order to take note of topics and concepts that might require further study.

Understanding the AP Biology Examination

You are probably aware that in general AP exams are long. The AP Biology Exam takes three hours. The exam probably looks like many other tests you've taken. It is made up of a multiple-choice section and a free-response (essay) section. At the core of the examination are questions designed to measure your knowledge and understanding of modern biology. You should be prepared to recall basic facts and concepts, to apply scientific facts and concepts to particular problems, to synthesize facts and concepts, and to demonstrate reasoning and analytical skills by organizing written answers to broad questions.

The AP Biology Exam is very challenging. When you sit down to take the test, exam administrators expect you not only to be fluent in the areas of biology that you find fascinating (the ones that probably inspired you to take a special interest in the subject originally), but they will expect you also to have an intimate knowledge of topics you don't find interesting at all. Whatever those topics might be—chemistry, DNA replication, the dizzying details of animal and plant classification, or cell organization—you need to be comfortable with and knowledgeable about all of the AP Biology topics.

Section I: Multiple-Choice Questions

Section I contains 120 multiple-choice questions that test both scientific facts and their applications. You will have 90 minutes to complete Section I. This portion of the exam is followed by a 5–10 minute break—the only official break during the examination. The directions for the multiple-choice section of the test are straightforward and similar to the following:

> **Directions:** Each of the questions or incomplete statements below is followed by five suggested answers or completions. Select the choice that best answers or completes the question or statement, and fill in the corresponding oval on the answer sheet.

It will probably not surprise you to read that not all multiple-choice questions are the same. In fact, the AP Biology Exam will contain the different types of multiple-choice questions listed and described below.

Factual Questions

The first type of multiple-choice question is your basic factual recall question, which will test whether or not you've memorized certain facts, processes, cycles, systems, etc. Here's an example of one of these:

1. In plants, which hormone is responsible for fruit ripening?
   (A) Auxin
   (B) Ethylene
   (C) Cytokinin
   (D) Phytochrome
   (E) Gibberellin

With a question like this, you either know it or you don't. This type of question is easy to read and quick to answer—that is, if you *remember* the correct answer! It's what the College Board calls a "factual" question: They ask a question, and you're expected to know the answer. The best way to approach factual questions is to read the question and every one of the five choices carefully. If you are certain you know the answer, fill in the corresponding oval on the answer sheet. However, what if you're not certain? The next step is to see if you can eliminate one or more of the choices.

Let's look again at the question above. Imagine that you don't recall that ethylene (answer choice *B*) is responsible for fruit ripening. You might realize that you can eliminate answer choice *D*, because the prefix *phyto-* suggests that the hormone phytochrome probably has something to do with plants' response to light and therefore phytochrome probably doesn't contribute to ripening. Now you have a one-in-four chance of making a correct guess. If you remember what any of the other three plant hormones do, your odds of guessing correctly increase even more. As a general statistical rule on AP exams, if you can eliminate two or more answer choices, you are better off making an educated guess than you are leaving the question unanswered. We'll discuss this idea further in the section "Grading Procedures for the AP Biology Examination."

"Reverse" Multiple-Choice Questions

Sometimes the College Board modifies the format for factual questions slightly, and the result is a slightly more difficult type of multiple-choice question. This type of question features four answers that are correct and only one that is incorrect; you are asked to find the *incorrect* choice. Throughout this book, these kinds of questions are identified as "reverse" multiple-choice questions; generally these questions contain the word *not* or the word *except*. Pay attention to the capitalization of these words in Section I of the exam to avoid making a careless mistake. Here is an example of what you might expect to see on the AP Biology Exam.

2. All of the following statements about photosynthesis are true EXCEPT
    (A) the light reactions convert solar energy to chemical energy in the form of ATP and NADPH
    (B) the Calvin cycle uses ATP and NADPH to convert CO2 to sugar
    (C) photosystem I contains P700 chlorophyll *a* molecules at the reaction center; photosystem II contains P680 molecules
    (D) in chemiosmosis, electron transport chains pump protons (H+) across a membrane from a region of high H+ concentration to a region of low H+ concentration
    (E) the steps of the Calvin cycle are sometimes referred to as the dark reactions because they do not require light in order to take place

This question asks you not only to remember one simple fact (like the function of a plant hormone), but also to consider the results of cellular processes and how systems in the cell compare. The correct answer here is *D*, because the statement in choice *D* is *not* true (in fact, the electron transport chains pump protons across membranes from regions of *low* H+ concentrations to regions of *high* H+ concentrations). Of course, the statements in choices *A, B, C,* and *E* are true. You may recall that this proton pumping occurs both in mitochondria and chloroplasts, and that the protons then diffuse—with the concentration gradient—back across the membrane (through ATP synthase), and that this drives the synthesis of ATP. However, you can also apply common sense to see that *D* doesn't look right. Why would a pump be needed to transport H+ ions *with* their concentration gradient? You may be pleased to know that this question is among the most difficult you will ever see on the AP Biology Exam. That is because photosynthesis is one of the most complicated processes in first-year biology.

Conceptual-Thematic Questions

Now let's look at another type of multiple-choice question that you'll encounter on the exam. The College Board calls these "conceptual-thematic" questions.

3. Which of the following groups is characterized by having a gastrovascular cavity, with a single opening acting as both mouth and anus, and existing in either polyp or medusa form?
   (A) Sponges
   (B) Cnidarians
   (C) Ctenophores
   (D) Platyhelminthes
   (E) Rotifers

Though it may be hard to see the difference between this question and a factual recall question, the difference is that this one asks you to use logic and synthesis to glean the sum of the organismal characteristics listed above. In other words, you're required to synthesize information rather than merely to recall a fact. In this case, the correct choice is *B*. The cnidarians you might encounter in everyday life are hydras, jellyfish, sea anemones, and coral animals. They are simple sac-like creatures with a gastrovascular cavity that contains a single opening. These invertebrates use tentacles to push prey from the water around them into their mouths. The food is then processed and ejected out the very same opening.

Matching Questions

Another type of multiple-choice question you'll see in Section I of the AP Biology Exam is the matching question. Below is an example of how matching questions are presented.

Questions 4–8
   (A) Savanna
   (B) Chaparral
   (C) Rainforest
   (D) Coniferous forest
   (E) Tundra

4. Characterized by epiphytes, closed canopy, and pronounced vertical stratification

5. Characterized by cone-bearing trees, dominated by one or two species of trees, and receiving heavy snowfall in the winter

6. Dominated by spiny evergreen shrubs, which are dependent on seasonal shrub fires for growth

7. Characterized as having insects as their dominant herbivores, predominant grass growth, and large grazing mammals

8. Characterized by permafrost, very low temperatures, and low annual rainfall

Your job is to match one correct choice to each of the numbered items. In general, matching questions are easier than factual multiple-choice questions, because knowing the answer to one question decreases the number of choices for the other four; knowing two further decreases the choices for the other three; and so on. To satisfy your curiosity, the correct answers above are 4. *C*; 5. *D*; 6. *B*; 7. *A*; and 8. *E*.

Lab-Based or Experimental Questions

The last type of multiple-choice question you'll see in Section I is the lab-based or experimental question. These questions either present you with a set of data in graph (or other) form, or they describe an experiment and ask you to make educated guesses and to form hypothesis. Take a look at the question below, for example. The graph shows the results of a study to determine the effect of soil air spaces on plant growth.

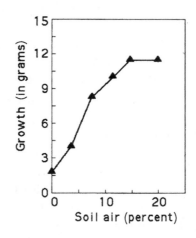

9. The data from the above graph shows that the plant
   (A) grows fastest when the soil is 5–10% air
   (B) grows fastest when the soil is 15–20% air
   (C) grows at the same rate no matter what the air percentage of the soil is
   (D) grows most slowly when the soil is 5–10% air
   (E) does not grow at all when the soil is 0–3% air

The correct choice is *A*. The graph shows the line with the greatest slope (the highest degree of change over the shortest amount of time) between the percentages 5 and 10. During this time, the plant grows by about 9 –5 = 4 grams. Just to be sure, check the amount this plant grows when the soil is 15–20% air. At the start, when the air was 15% air, the plant weighed 12 grams. At the end, when the soil was 20% air, the plant weight the same—12 grams. Virtually no growth occurred during this time at all.

Clearly this question requires you to be able to read a graph, but at this point in your biology education you should be quite capable of doing that. In order to brush up on the various ways that graphs present information, you might spend some time looking over Chapter 23 of Campbell and Reece *Biology*.

Section II: Free-Response Questions

Section II of the AP Biology Exam is made up of four free-response questions relating to the following topics: molecules and cells, heredity and evolution, and organisms and populations (two questions). One or more of the four free-response questions will be lab-based, since the test writers expect you to be knowledgeable with laboratory concepts and techniques. Many of the questions you encounter on Section II of the AP Biology Exam will require you to integrate material from across the topic outline. The free-response questions are often broken down into parts, and the parts vary in difficulty. Presenting free-response questions in this form is the College Board's way of making sure that you really understand the underlying concepts of biology—and that you aren't just a really lucky guesser.

At the beginning of Section II of the AP Biology Exam, you'll be given a green booklet in which to write your essays. Here's what you'll see when you open your test booklet to Part II.

BIOLOGY

Section II

Time—1 hour and 30 minutes

Answer all questions. Number your answer as the question is numbered below.

Answers must be in essay form. Outline form is NOT acceptable. Labeled diagrams may be used to supplement discussion, but in no case will a diagram alone suffice. It is important that you read each question completely before you begin to write.

1. Water comprises roughly 70% of the human body; cells are roughly 70–95% water, and water covers about three-quarters of the Earth's surface.
   - **Describe** the major physical properties of water that make it unique from other liquids.
   - **Explain** the properties of water that enable it to travel up through the root and stems of plants to reach the leaves.

▌**Explain** why the temperature of the oceans can remain relatively stable and support vast quantities of both plant and animal life, when air temperature fluctuates so significantly throughout the year.

Like many free-response questions on the AP Biology Exam, this one is broken into three distinct parts. Each contains a clear directive. In fact, they are printed in boldface to help you focus on exactly how you should answer the question. First you will need to explain the uniqueness of water by describing its major physical properties. (In your response to this first part of the question, you might wish to include a diagram of the structure of water, complete with electrons and bonds.) Then you must explain the properties of water that allow it to travel from root to leaf. Finally, you should explain the reason(s) why ocean water temperature remains stable and supports plant and animal life—even in the face of great air temperature variations. Of course, limiting your answer by addressing *exactly* what the question asks will make writing the essay easier for you and earn you a higher score. Always take the time to determine precisely what is being asked before you begin to formulate a concrete thesis and focus on writing your relevant supporting paragraphs.

Grading Procedures for the AP Biology Examination

The raw scores of the AP Biology Examination are converted to the following 5-point scale:

5—Extremely Well Qualified
4—Well Qualified
3—Qualified
2—Possibly Qualified
1—No Recommendation

Some colleges give undergraduate course credit to students who achieve scores of 3 or better on AP exams. Other colleges require students to achieve scores of 4 or 5. Remember, if you are considering using your AP Biology Exam score for college credit, you should check with individual colleges to find out their specific requirements for credit. Below is a breakdown of how the grading of the AP Biology Exam works.

Section I: Multiple-Choice Questions

The multiple-choice section of the exam is worth 60% of your total score. The raw score of Section I is determined by crediting one point for each correctly-answered question and by deducting ¼ point for each question answered incorrectly. No points are gained or lost for unanswered questions. Consequently, if you are able to eliminate two or more of the five answer choices, statistically it is to your advantage to make an educated guess rather than to leave the answer blank. Naturally it follows that the more answer choices you can eliminate, the better your chance of guessing the right answer.

Section II: Free-Response Questions

Section II counts for 40% of your examination grade. Within Section II, each of the four essay responses is weighted equally. While the multiple-choice section of the exam is graded by tirelessly perfect computers that will ensure you're exactly compensated for your efforts, the free-response section is scored by actual high school and college faculty members, referred to as "readers", who are carefully chosen on the basis of their teaching experience, expertise, gender, ethnicity, and geographic location. The College Board has tried to ensure that scoring is done with as little bias as possible. Toward this end, it has developed guidelines to ensure that all readers use exactly the same guidelines for scoring free-response questions. Moreover, a different reader will score each of your four essays, and each of your scores (as well as your name and school information) will be covered up as each reader evaluates an essay. These procedures are used to guard against readers giving better or worse scores to students based on how well they have performed on other questions—or on personal or geographical information.

Your answers to the free-response questions must be presented in essay form. The College Board does not give credit for outlines or for unlabeled and unexplained diagrams. For the free-response section, each of the four responses is judged on an 11-point scale, which goes from 0–10. Each free-response question carries its own scoring guidelines that are specific to the subject matter of the question.

Test-Taking Strategies for the AP Biology Examination

Here are a few tips for preparing yourself in the weeks leading up to the examination.

▌ The earlier you start studying for the AP Biology Exam, the better. Some people believe in using the AP Biology prep book along with their textbook throughout the course, taking notes in the margin to supplement their teacher's lectures. You should definitely begin to prepare for the test at least one month in advance. Spend an hour a day (or perhaps two hours every *other* day) making your way through Part II of this book, rereading your lecture notes, and reading the corresponding sections in the Campbell and Reece textbook *Biology* to study a topic in more depth. You can use the correlation guide at the end of Part I of this book to link AP Biology topics to your textbook.

▌ Review lab materials from whatever sources you have. If you have a published lab manual, you might read through all the labs in it. If your teacher is using a lab manual that is above high school level, you might ask him or her to point to the labs that might be covered on the exam. If your teacher photocopies loose papers for you, organize and study them. The College Board suggests the following twelve labs for AP Biology courses, so you should be familiar with them. (You can find more detailed descriptions of these labs in the AP Biology *Course Description* or on the College Board's website.)

1. Diffusion and Osmosis
2. Enzyme Catalysis
3. Mitosis and Meiosis
4. Plant Pigments and Photosynthesis
5. Cell Respiration
6. Molecular Biology
7. Genetics of Organisms
8. Population Genetics and Evolution
9. Transpiration
10. Physiology of the Circulatory System
11. Animal Behavior
12. Dissolved Oxygen and Aquatic Primary Productivity

▌ As you study, you might benefit from using mnemonics wherever you can. Mnemonics is a method of remembering a set of terms in a certain order. For instance, look at the stages of mitosis:

Prophase
Metaphase
Anaphase
Telophase
Cytokinesis

Having to remember those phases in the correct order is a challenge. Well, what if all you had to remember was "**Persuade My Aunt To Conga**"—or some other funny or memorable sentence that uses the first letters of these five words? If you are familiar with the mitotic material in your textbook, this kind of prompting can help you recall the five stages quickly. This method works well for ordered lists of terms (and there are plenty of these in AP Biology).

▌ After you've done a good amount of studying, you should take the first sample exam in Part III of this book and see how you do. Circle the items you get wrong and list the subject matter that those questions involve. Then analyze the list to see if there are patterns—are you missing mostly questions on animal physiology or the process of photosynthesis? Spend the next week or so studying the topics in which you are weak. Then, perhaps a week or two before the AP Biology Exam, take the second sample exam. Once again, grade this exam and see where your weak points are. Then you can spend the final days before the test looking through this guide, your class notes, and your textbook to fill in any remaining gaps.

▌ When you take the sample tests, approximate as closely as you can the actual conditions of the exam site. In other words, leave yourself three hours and ten minutes, sit at a desk or table, make certain that you won't be interrupted by the phone, or distracted by the TV, stereo, or other people. This will help you develop the focus and stamina necessary to do well on the AP Biology Exam.

▌ Finally, it's probably a good idea to put aside studying the night before the exam. If you've been preparing for weeks beforehand, you're ready to take the test. Instead, your goal is to try to relax.

Below is a brief list of basic tips and strategies to think about *before* you arrive at the exam site.

▌ It's a good idea to arrive at the exam site thirty minutes before the start time. This saves you anxiety about arriving late. Plan your schedule so that you get *two* very good uninterrupted nights of sleep before exam day. On the day of the exam, make sure that you eat a good, nutritious meal. These tips may sound corny or obvious, but your body must be in peak form in order for your brain to perform well.

▌ Have a photo I.D. with you when you arrive at the exam site. (It's essential if you are taking the exam at a school other than your own.) Carrying a driver's license or a student I.D. card will allow you to prove your identity.

▌ Bring at least two sharpened #2 pencils for the multiple-choice section, as well as two black or dark blue pens for the free-response section of the exam. Also, bring a clean pencil eraser with you. The machine that scores Section I of the exam recognizes only marks made by a #2 pencil. Also, it cannot read a correct answer if a previous answer has not been erased completely.

▌ If possible, try to bring a watch with you to the exam. Most testing rooms do have clocks. Still, having your own watch makes it easy to keep close track of your own pace. The watch cannot have a calculator or an alarm, however, as these are not permitted in the exam room.

▌Here are a few other things that are not allowed in the exam room:

1. books
2. laptop computers
3. beepers
4. cameras
5. portable radios

If you must bring a cellular phone with you, be prepared to turn it off and to give it to the test proctor until you are finished with your exam.

Educational Testing Service prohibits the objects listed above in the interest of fairness to all test-takers. Similarly, the test administrators are very clear and very serious about what types of conduct are not allowed during the examination. Below is a list of actions to avoid at all costs, since each is grounds for your immediate dismissal from the exam room.

▌Do not consult any outside materials during the three hours of the exam period. Remember, the break is technically part of the exam—you are not free to review any materials at that time either.

▌Do not speak during the exam, unless you have a question for the test proctor. Raise your hand to get the proctor's attention.

▌When you are told to stop working on a section of the exam, you must stop *immediately*.

▌Do not open your exam booklet before the test begins.

▌Never tear a page out of your test booklet or try to remove the exam from the test room.

▌Do not behave disruptively—even if you're distressed about a difficult test question or because you've run out of time. Stay calm and make no unnecessary noise. Remember, too, the worst-case scenario: if you are displeased with your performance on test day, you can always cancel your exam scores.

Section I: Strategies for Multiple-Choice Questions

Obviously, having a firm grasp of biology is, of course, the key to doing well on the AP Biology Examination. In addition, being well-informed about the exam itself increases your chances of achieving a high score. Below is a list of strategies that you can use to increase your comfort, your confidence, and your chances of excelling on the multiple-choice section of the exam.

▌Pace yourself and keep track of the remaining time as you complete the multiple-choice section of the exam. Remember, you have ninety minutes to answer all 120 questions (that is about 45 seconds per question). It's important that you don't get stuck on one question for too long.

▌Make a light mark in your test booklet next to any questions you can't answer. Return to them after you reach the end of Section I. Sometimes questions that appear later in the test will refresh your memory on a particular topic, and you will be able to answer one of those earlier questions.

- Always read the entire question carefully, and underline key words or ideas. You might wish to double underline words such as *NOT* or *EXCEPT* in that type of multiple-choice question.
- Read each and every one of the answer choices carefully before you make your final selection.
- Trust your first instinct. It has been proven statistically that your first choice is more likely to be correct, so you should replace your first choice only if you are completely certain that your second choice is correct.
- Use the process of elimination to help you arrive at the correct answer. Even if you are quite sure of an answer, cross out the letters of incorrect choices in your test booklet as you eliminate them. This cuts down on the incorrect choices and allows you to narrow the remaining choices even further.
- If you are able to eliminate two or more answer choices, it is better to make an educated guess at the correct answer than to leave the answer blank.
- Try to grow as familiar as you can with the format of Section I. The more comfortable you are with the multiple-choice format and with the kinds of questions you'll encounter, the easier the exam will be. Remember, Part II and Part III of this book provide you with invaluable practice on the kinds of multiple-choice questions you will encounter on the AP Biology Exam. Once again, here are the five types of multiple-choice questions you'll see on the exam:

  - Factual recall questions;
  - "Reverse" multiple-choice questions;
  - Conceptual-Thematic questions;
  - Matching questions;
  - Lab-based or Experimental questions.

- Become completely familiar with the instructions for the multiple-choice questions *before* you take the exam. You'll find the instructions in this book. By knowing the instructions cold, you'll save yourself the time of reading them carefully on exam day.

Section II: Strategies for Free-Response Questions

Below is a list of strategies that you can use to increase your chances of excelling on the free-response section of the exam.

- Since you have just ninety minutes to outline and write four essays in the free-response section of the AP Biology Exam, you must manage your time carefully. You have about twenty-two minutes to spend on each essay.
- Read the first question and quickly jot down any phrases, terms, or other details that you remember about the topic. This process can jog your memory further, and it allows you to take advantage of what might be called your "first-glance recall."
- Then read the question again and determine exactly what is being asked. As you read a question, underline any directive words (usually the first word in an essay) that indicate how you should answer and focus the material in your

essay. Some of the most frequently used directives on the AP Biology Exam are listed below, along with descriptions of what you need to do in your writing to answer the question.

- *Analyze* (show relationships between events; explain)
- *Compare* (discuss similarities between two or more things)
- *Contrast* (discuss points of difference or divergence between two or more things)
- *Describe* (give a detailed account)
- *Design* (create an experiment and convey its ideas)
- *Explain* (clarify; tell the meaning)

▌ Once you've determined what you remember about a topic and analyzed exactly what is being asked, you should create a thesis, or plan, that will allow you to use as many facts as possible in your response to the question. As you formulate your thesis, always consider whether or not it will answer the essay question directly.

▌ Make a quick outline of your approach to the question. Reread the question as many times as necessary to make sure that you will cover each aspect of the topic. Free-response questions frequently have several parts, so you will need to take this into account as your outline your ideas.

▌ As the exam states clearly in the directions to free-response questions, a diagram or graph by itself is *never* an acceptable way to answer a free-response question. However, you should think about whether you can use a labeled diagram or graph (*in addition* to your written response) to develop your answer in some useful way. Don't bother to include a graph if it doesn't accomplish this goal.

▌ Each of your essays should include an introductory paragraph that asserts a thesis, paragraphs that support your argument, and a concluding paragraph that summarizes your argument.

▌ Always use concrete scientific ideas and examples to substantiate your thesis, and avoid including any information that you are not certain is correct.

▌ If time allows, proofread your essays. Don't worry about crossing out material—readers understand that your responses are first drafts and that you are writing down ideas under the pressure of time. Your answers should be legible, but they do not need to reflect perfect penmanship. Focus your efforts on making the essays strong by backing up your thesis with a clear organization and plenty of scientific detail.

The success of your four free-response essays will depend a great deal on how clearly and extensively you answer the question being posed. Of course, the structure of your essays will depend entirely on your knowledge of the subjects at hand. In the introduction of your essay, you should indicate how you will respond to the question. This is the point when you can address the general topic of the question. Take a look at an example of another free-response question below.

2. In cancer, the cell's reproductive machinery experiences a loss of control that makes cancer cells reproduce continually, which eventually form tumors.

▌ **Describe** three DNA-related cellular events that could lead to the loss of cell

division control that causes cancer.
- **Describe** why tumors are detrimental to the body.
- **Discuss** several cell processes that you think should be studied more closely, in finding a cure for cancer.

In order to answer this question, you should isolate exactly what it is that you must answer. You may want to underline the relevant information in the question to remind yourself of your focus:

2. In cancer, the <u>cell's reproductive machinery</u> experiences a loss of control that makes <u>cancer cells reproduce</u> continually, which eventually <u>form tumors</u>.
- **Describe** <u>three DNA-related cellular events</u> that could <u>lead to the loss of cell division control</u> that <u>causes cancer</u>.
- **Describe** why <u>tumors</u> are <u>detrimental</u> to the body.
- **Discuss** <u>several cell processes</u> that you think should be <u>studied more closely</u>, in finding a <u>cure for cancer</u>.

In order to answer the first part of the question, you'll need to identify three appropriate DNA-related cellular events.

1. An error in the proofreading of DNA
2. Errors in DNA replication that go undetected by the cell's proofreading devices
3. A translocation

Under each of the three events, you should list any and all details you remember about those events to use in your description. When you flesh out these details, you'll need to clearly connect them to the concept of the cancer-causing loss of cell division control.

To answer the second part of the question, you'll need to list as many reasons as you can find why tumors are harmful to the body. These might include cancerous cells' breaking off, entering the bloodstream, ending up elsewhere in the body, and continuing to divide to form another tumor; tumors' ability to occur anywhere in the body; their tendency to block the flow of blood when they grow near blood vessels; disrupting the natural function of any organ in the body; and endangering homeostasis. Of course, after you list reasons, you'll need to add details to each item in your list.

To answer the final part of the question, you'll need to consider first what causes cancer. Then you must think creatively in order to suggest possible approaches to dealing with each specific cause.

The last paragraph of your essay should synthesize your responses to the question's three parts and offer an overall conclusion.

In the next section of this book, we'll do a review of everything that you learned in your textbook that will be on the AP Biology test, and we'll pose many questions along the way so that you can get used to being tested on the concepts in the way that the College Board will test you. At the end of the book, in Part III, there are two practice tests for you to try on your own, along with complete answers.

AP Topic Correlation to Campbell and Reece *Biology,* Sixth Edition

The following chart is intended to help you study for the AP Biology Exam. The left column includes a series of AP Biology topics with which you should be familiar before you take the AP Biology Exam. The right column includes a detailed breakdown of chapters in your Campbell and Reece *Biology,* Sixth Edition, textbook where you can learn more about those topics. This guide is also useful with other editions of your textbooks, although some illustrations and page or chapter numbers may have changed. You may want to use this chart throughout the year to review what you've learned. It is also an excellent place to begin your pre-exam review of subjects.

AP BIOLOGY TOPICS	CORRELATIONS TO CAMPBELL AND REECE BIOLOGY *(Sixth Edition)*
MOLECULES AND CELLS: CHEMISTRY OF LIFE	
Water	**Chapter 3: Water and the Fitness of the Environment** The Effects of Water's Polarity The Dissociation of Water Molecules
Organic Molecules in Organisms	**Chapter 5 The Structure and Function of Macromolecules** Polymer Principles Carbohydrates—Fuel and Building Material Lipids—Diverse Hydrophobic Molecules Proteins—Many Structures, Many Functions Nucleic Acids—Informational Polymers
Free Energy Changes	**Chapter 6: An Introduction to Metabolism** Metabolism, Energy, and Life Enzymes The Control of Metabolism
Enzymes	**Chapter 6: An Introduction to Metabolism** Enzymes The Control of Metabolism
MOLECULES AND CELLS: CELLS	
Prokaryotic and Eukaryotic Cells	**Chapter 7: A Tour of the Cell** How We Study Cells A Panoramic View of the Cell The Nucleus and Ribosomes
Membranes	**Chapter 7: A Tour of the Cell** A Panoramic View of the Cell Cell Surfaces and Junctions
Subcellular Organization	**Chapter 7: A Tour of the Cell** A Panoramic View of the Cell The Nucleus and Ribosomes The Endomembrane System

Mutation	**Chapter 15: The Chromosomal Basis of Inheritance** Errors and Exceptions in Chromosomal Inheritance
	Chapter 17: From Gene to Protein The Synthesis of Protein
Viral Structure and Replication	**Chapter 18: Microbial Models: The Genetics of Viruses and Bacteria** The Genetics of Viruses
Nucleic Acid Technology and Applications	**Chapter 20: DNA Technology and Genomics** DNA Cloning DNA Analysis and Genomics Practical Applications of DNA Technology

HEREDITY AND EVOLUTION: EVOLUTIONARY BIOLOGY

Early Evolution of Life	**Chapter 26: Early Earth and the Origin of Life** Introduction to the History of Life The Origin of Life The Major Lineages of Life
Evidence for Evolution	**Chapter 25: Phylogeny and Systematics** The Fossil Record and Geologic Time
Mechanisms of Evolution	**Chapter 22: Descent with Modification: A Darwinian View of Life** The Historical Context for Evolutionary Theory The Darwinian Revolution
	Chapter 23: The Evolution of Populations Population Genetics Causes of Microevolution Genetic Variation, the Substrate for Natural Selection
	Chapter 24: The Origin of Species What Is a Species? Modes of Speciation From Speciation to Macroevolution
Mechanisms of Evolution (*continued*)	**Chapter 25: Phylogeny and Systematics** The Fossil Record and Geologic Time Systematics: Connecting Classification to Phylogeny

ORGANISMS AND POPULATIONS: DIVERSITY OF ORGANISMS

Evolutionary Patterns	**Chapter 23: The Evolution of Populations** Population Genetics Causes of Microevolution Genetic Variation, the Substrate for Natural Selection
Survey of the Diversity of Life	**Chapter 25: Phylogeny and Systematics** The Fossil Record and Geologic Time Systematics: Connecting Classification to Phylogeny
	Chapter 26: Early Earth and the Origin of Life Introduction to the Early History of Life The Origin of Life The Major Lineages of Life

ORGANISMS AND POPULATIONS: STRUCTURE AND FUNCTION OF PLANTS AND ANIMALS

ORGANISMS AND POPULATIONS: ECOLOGY

Part II

Topical Review with Sample Questions and Answers and Explanations

Part II is keyed to *Biology,* Sixth Edition, by Campbell and Reece. It gives an overview of important information in bulleted form and provides sample multiple-choice and free-response questions, along with answers and explanations. Use the topical review and the Summary of Key Concepts sections at the end of each chapter in your textbook before attempting the practice questions. Be sure to review the answers thoroughly to prepare yourself for the range of test items you will encounter on the AP Biology Examination.

CHAPTER 1

The Chemistry of Life

The Chemical Context of Life: Chemical Elements and Compounds

▌ **Atoms** are the smallest unit of an element that still retains the property of the element. Atoms are made up of neutrons, protons, and electrons.

▌ **Neutrons** and **protons** are close together in the nucleus of the atom, and **electrons** move quickly in a cloud around the nucleus.

▌ The number of protons an element possesses is referred to as its **atomic number**, and this number is unique to every element. The **mass number** of an element is the number of protons and neutrons.

Atoms and Molecules

▌ **Chemical bonds** are defined as interactions between the valence electrons of atoms, and atoms are held together by chemical bonds to form **molecules**.

▌ A **covalent bond** occurs when valence electrons are shared by two atoms. **Nonpolar covalent bonds** occur when the electrons being shared are shared equally between the two atoms. In **polar covalent bonds**, the two atoms have different electronegativities, resulting in an unequal sharing of the electrons.

▌ **Ionic bonds** are ones in which the two bonded atoms attract the shared electrons so unequally that the more electronegative atom steals the electron away from the less electronegative atom. Ionic bonds form ionic compounds or salts.

▌ **Hydrogen bonds** are relatively weak bonds that form between molecules, as in water. In hydrogen bonds, the positively charged hydrogen atom of one molecule is attracted to the negatively charged atom of another molecule.

▌ **Organic molecules** have different properties as a result of their different structures. More specifically, the behavior of organic molecules is dependent on the identity of their functional groups.

▌ Some common functional groups are listed below:

Functional Group	Organic Molecules with the Functional Group	Chemical Properties of Functional Group
Hydroxyl Group, —OH	alcohols such as ethanol, methanol, etc.	hydrophilic and polar in nature
Carboxyl, —COOH	carboxylic acids such as fatty acids and sugars	hydrophilic and polar
Carbonyl, —COR or —COH	ketones, and aldehydes such as sugars	hydrophilic and polar
Amino Group, —NH_2	amines such as amino acids	hydrophilic and polar
Phosphate Group, PO_3	organic phosphates, including ATP, DNA, and phospholipids	hydrophilic and polar
Methyl Group, —CH_3	found in butane	hydrophobic

Water and the Fitness of the Environment: The Effects of Water's Polarity

▌ The **structure of water** is the key to its special properties. Water is made up of one atom of oxygen and two atoms of hydrogen, bonded to form a molecule that looks like this:

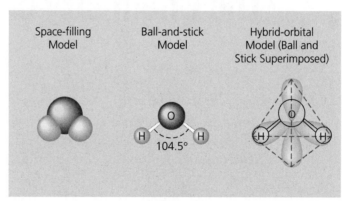

Water (H$_2$O)

▌ The fact that water is **V-shaped** means that it is polar (the opposite ends of the molecule have opposite charges). The end bearing the oxygen atom has a slight negative charge, whereas the end bearing the hydrogen atoms has a slightly positive charge. This is what makes water able to form **hydrogen bonds**—the slightly negative oxygen atom from one water molecule is attracted to the slightly positive hydrogen end of another water molecule. Each water molecule can form a maximum of four hydrogen bonds at a time.

▌ **Adhesion** is the clinging of one substance to a water molecule. Water is very adhesive because of its hydrogen bonds. It is this property of adhesion that allows water to travel up the stems of plants. In **transpiration**, water evaporates from the leaves of plants. Since hydrogen bonding virtually attaches the water molecules one to another to form a long chain up through the cells of the plants, the water is pulled up in a chain through the xylem by **capillary action**. Because of water's **cohesion**, water molecules have the ability to cling to each other, as well.

▌ Water is responsible for insulating the earth, and **moderating its temperature**. It is able to do this because of its very high specific heat. **Specific heat** is the amount of heat required to raise or lower the temperature of a substance by 1 degree Celsius, and the specific heat of water is 1 cal/g·°C. (Water has very high specific heat, so it takes more heat energy to raise the temperature of water, and more energy must be lost from water before its temperature will decrease.) The reason for water's high specific heat is its hydrogen bonds. But this high specific heat makes the temperature of the earth's oceans relatively stable and able to support vast quantities of both plant and animal life.

▌ Water's **solid state** is less dense than its **liquid state**, whereas the opposite is true of most other substances. This is because solid water forms a regular **crystal lattice** structure, in which each water molecule is hydrogen bonded to four other water molecules. The fact that ice is less dense than water enables ice to float on ponds and lakes and ice cubes to float in your glass of soda.

█ Water is also an important **solvent**. (You may recall that the substance that something is dissolved in is the solvent, while the substance being dissolved is the **solute**. Together they are the **solution**.) When the solution has water as its solvent, it's called an **aqueous solution**. Substances that are water-soluble are **hydrophilic** substances. They are ionic compounds, polar molecules (i.e. sugars), and some proteins. Oils, however, are **hydrophobic** and non-polar, meaning that they do not dissolve in water.

The Dissociation of Water Molecules

When a hydrogen atom is transferred from one water molecule to another, it leaves its electron and is transferred as a **hydrogen ion**, which is a proton with a charge of +1. This transfer makes the water molecule that lost its proton the **hydroxide ion** (depicted as OH^-) and the molecule that gains the proton is **hydronium ion**, H_3O^+. This **dissociation** of water molecules happens such that the amount of hydronium and hydroxide ions is about even in pure water. But if acids or bases are added to water, this equilibrium shifts. Water has a **pH** of 7, which means it is neutral.

The Structure and Function of Macromolecules: Polymer Principles

█ Most large organic molecules (macromolecules) are **polymers**. Polymers are long chain molecules made up of repeating units that are either the same as—or very similar to—each other. The small units that make up polymers are called **monomers**. What distinguishes polymers is the different identity of their monomers.

█ The reaction that creates polymers from monomers is called a **condensation**, or **dehydration reaction**. In this reaction, two monomers are combined, and one water molecule is released. The reverse reaction, in which a polymer is broken down into monomers after the addition of water, is called **hydrolysis**.

Carbohydrates—Fuel and Building Material

There are four basic kinds of macromolecules that are biologically important: carbohydrates, lipids, proteins, and nucleic acids. The term *carbohydrates* refers both to simple sugars (glucose, fructose, maltose, etc.) and to the polymers made from these and other subunits.

█ The simplest sugars are **monosaccharides**. Monosaccharides are simple ring structures like **glucose** and **fructose**. Both glucose and fructose have the same chemical formula $C_6H_{12}O_6$, but the way their atoms are arranged gives each one a different molecular formula and different chemical properties.

█ **Disaccharides** are made up of two monosaccharides that have undergone a condensation reaction. Three examples of these are **sucrose, maltose,** and **lactose**.

█ **Polysaccharides** are basically polymers of monosaccharides. Polysaccharides are involved in the storage of carbohydrates in organisms (both plant and animal). In plants, carbohydrates are stored in the form of **starch**, which is made up of glucose monomers. In animals, carbohydrates are stored as **glycogen**.

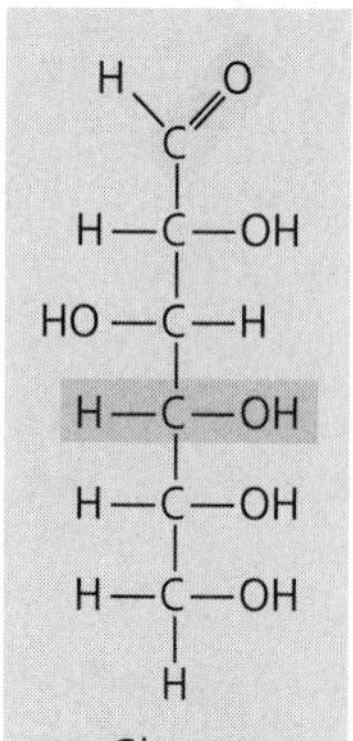

Glucose

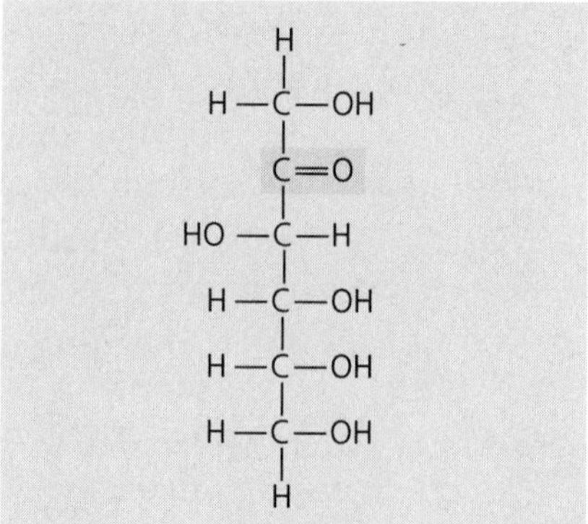

Fructose

This, too, is a polysaccharide made up of glucose, but its structure is more branched than is the structure of starch.

▌ Polysaccharides are also involved in the structure of organisms. **Cellulose** (a polysaccharide) makes up the thick cell walls of plants. Cellulose is a polymer of glucose, but again its identity differs from that of starch and glycogen because of the way the glucose molecules are joined.

▌ In animals such as arthropods (lobsters, crabs, other crustaceans, and insects), **chitin** is an important structural polysaccharide. Chitin is made up of a variation of glucose with a nitrogenous arm.

Lipids—Diverse Hydrophobic Molecules

Unlike the other three major groups of biological molecules, lipids aren't polymers. They are grouped together because they are hydrophobic, and this hydrophobic nature is based on their structure. There are many different kinds of lipids, some of which are **waxes**, **oils**, **fats**, and **steroids**.

▌ **Fats** are large molecules that are created by dehydration synthesis reactions between smaller molecules. Fats (also called triacylglycerols or triglycerides) are made up of a **glycerol** molecule and three **fatty acid** molecules.

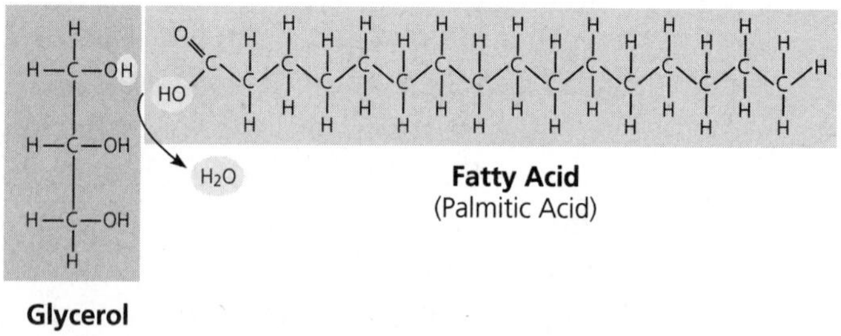

Glycerol

Fatty Acid
(Palmitic Acid)

Dehydration Synthesis

▌ Fats differ in the length of their hydrocarbon backbones and also in the presence and positions of the double bonds they contain between carbon atoms. **Saturated fatty acids** contain no double bonds, **unsaturated fatty acids** contain at least one double bond, and polyunsaturated fatty acids contain two or more double bonds.

- In animals, fat is an important storage molecule. In humans and other mammals, fat is stored in **adipose cells** (fat cells).
- Lipids are a very important part of **phospholipids**, which make up cell membranes. Phospholipids have a glycerol backbone and two fatty acid tails; the glycerol head is hydrophilic, and the fatty acid tails are hydrophobic. In forming the cell membrane, they are arranged in a bilayer, with their hydrophobic ends sandwiched in between the other portion of the cell membrane and the inner portion. The hydrophilic ends are pointing toward the cytosol and the outer cell environment.
- The final type of lipid you'll need to be familiar with is steroids. **Steroids** are made up of four rings that are fused together.

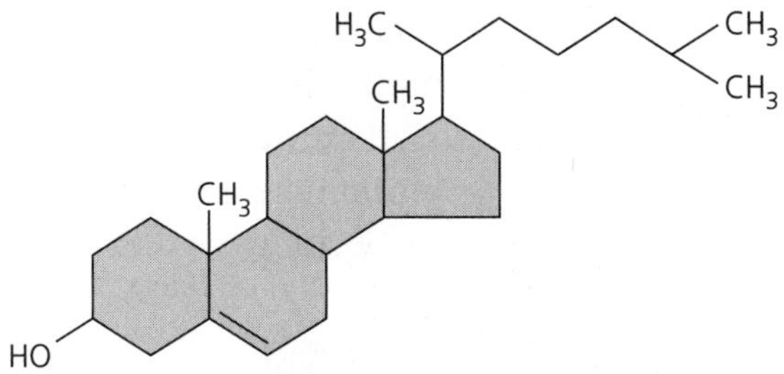

Cholesterol, a Steroid

One common type of steroid is cholesterol, shown above. Cholesterol is an important component of cell membranes. Another example is found in hormones such as estrogen and testosterone.

Proteins—Many Structures, Many Functions

Proteins are a very important component of the cell; in fact, they make up about 50% of the cell.

- **Proteins** are polymers made up of amino acid monomers.
- **Amino acids** are organic molecules that contain a carboxyl group and an amino group, as well as an R group (variable group), that gives each amino acid its identity and properties. There are twenty amino acids that make up protein molecules: glycine, alanine, valine, leucine, isoleucine, methionine, phenylalanine, tryptophan, proline, serine, threonine, cysteine, tyrosine, asparagine, glutamine, aspartic acid, glutamic acid, lysine, arginine, and histidine. You should be able to recognize from their names that they are amino acids, because most amino acids end in –*ine*.
- In proteins, amino acids are joined by **peptide bonds** in a dehydration synthesis reaction. The function of proteins depends on how many amino acids and what type of amino acids are joined together.
- There are four levels of protein structure. The most basic is the **primary structure**—the sequence in which the amino acids are joined.
- The **secondary structure** refers to one of two three-dimensional shapes that the protein can have due to its hydrogen bonding. One shape is a coiled shape called an alpha helix, and the second shape is an accordion shape called a beta pleated sheet.

- The **tertiary structure** of a protein refers to interactions between side chains of the protein. These interactions involve hydrophobic interactions, van der Waals forces, and disulfide bridges.
- The **quaternary structure** of a protein refers to the association of two or more polypeptide chains into one giant macromolecule, or functional protein.
- When a protein is **denatured**, upon heating or the introduction of a pH change or other disturbance, it becomes inactive. Denaturation causes the protein to lose its shape, or conformation.

Nucleic Acids—Informational Polymers

The last group of important biological molecules we'll discuss is the nucleic acids. The two nucleic acids are **DNA** (deoxyribonucleic acid) and **RNA** (ribonucleic acid).

- DNA is the molecule of heredity. It is what is inherited from cell to cell, parent organism to offspring. DNA molecules are very long; they are polymers of **nucleotide** monomers. Nucleotides are made up of three parts: a **nitrogenous base**, a five-carbon sugar called a **pentose**, and a phosphate group

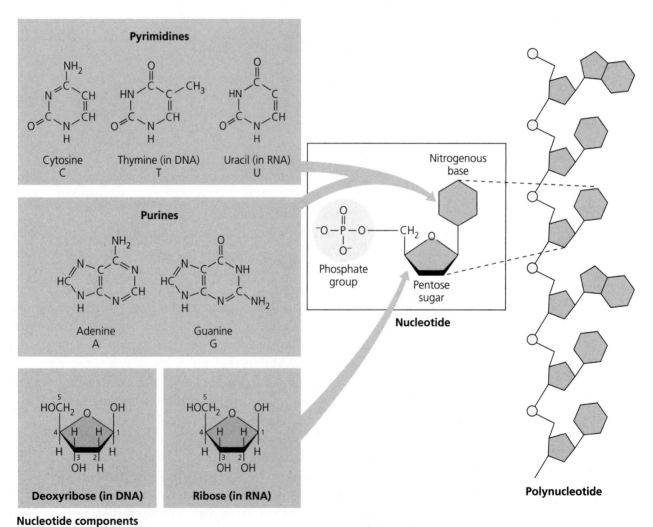

The Components of Nucleic Acids

- There are two types of nitrogenous bases, **purines** and **pyrimidines**. The purines are **adenine**, abbreviated A, and **guanine** (G), and the pyrimidines are **cytosine** (C), **thymine** (T) and **uracil** (U). Thymine is found only in DNA, and uracil is found only in RNA. In DNA, adenine always pairs with thymine, and cytosine always pairs with guanine.
- In DNA, the pentose sugar is **deoxyribose**, and in RNA the pentose sugar is **ribose**. Deoxyribose has one less oxygen than ribose.
- In DNA and RNA, the nucleotides are joined by **phosphodiester bonds**. In DNA, two polynucleotide chains wrap around each other in a helical shape, whereas RNA is a single polynucleotide.

An Introduction to Metabolism

The following information concerns how energy is used, how it moves, how it is released, and the laws that govern it.

Metabolism, Energy, and Life

- **Energy** is defined as the capacity to do work. Things that move are said to possess **kinetic energy**. An object at rest can possess **potential energy** if it has stored energy as a result of its position or structure. **Chemical energy** is stored in molecules, and the amount of chemical energy a molecule possesses depends on its chemical bonds.
- The study of energy transformations that occur in matter is called **thermodynamics**, and there are two basic laws of thermodynamics. The **first law of thermodynamics** states that the energy of the universe is constant and that energy can be transferred and transformed, though it cannot be created or destroyed. The **second law of thermodynamics** states that every energy transfer or transformation increases the **entropy**, or the amount of disorder or randomness, in the universe.
- **Free energy** is defined as the part of a system's energy that is able to perform work when the temperature of a system is uniform. Free energy is depicted as **G**. The symbol for the total energy of a system is **H**, and the symbol for entropy is **S**. The relationship between the change in free energy, change in energy, change in entropy, and temperature is as follows: $\Delta G = \Delta H - T\Delta S$.
- In order for a chemical reaction to occur spontaneously, the system must experience either a decrease in H (energy) or an increase in S (entropy).
- An **exergonic reaction** is one in which energy is released (ΔG is negative). An **endergonic reaction** is one that requires energy in order to proceed. Endergonic reactions absorb free energy (ΔG is positive).
- **ATP** is a very important molecule, because it is the primary source of energy for the cell. Also known as adenosine triphosphate, ATP is made up of the nitrogenous base adenine, bonded to ribose and a chain of three phosphate groups. When a phosphate group is hydrolyzed, energy is released in an exergonic reaction.
- Work in the cell is done by the release of a phosphate group from ATP. When ATP transfers one phosphate group through hydrolysis, it becomes **ADP** (**adenine diphosphate**).

■ The ATP cycle in the cell involves the use of energy that's released from catabolic reactions to re-phosphorylate ADP to form ATP. Energy is released when ATP is then dephosphorylated in an exergonic reaction to power important cellular processes.

Enzymes

■ **Catalysts** are substances that can change the rate of a reaction without being altered themselves in the process. **Enzymes** are proteins that are biological catalysts.

■ The **activation energy** of a reaction is the amount of energy it takes to start a reaction—the amount of energy it takes to break the bonds in the reactant molecules.

■ Enzymes speed up reactions by lowering the activation energy of the reaction—but without changing the free energy change of the reaction. The reactant that the enzyme acts on is called a **substrate**.

■ A certain region in the enzyme, known as the **active site**, is the part of the enzyme that binds to the substrate. The enzyme and substrate forms a complex called an **enzyme-substrate complex** that is held together by weak interactions. The substrate is then converted into **products**, and the products are released from the enzyme.

■ There are some substances that inhibit the actions of enzymes. **Competitive inhibitors** are reversible inhibitors that compete with the substrate for the active site on the enzyme. **Noncompetitive inhibitors** bind to another site on the enzyme, other than the active site; this causes the enzyme to change its shape, preventing the substrate from binding to the active site.

■ Many enzyme regulators bind to an allosteric site on the enzyme, which is a specific receptor far from the active site. Once bound, they can either stimulate or inhibit enzyme activity.

For Additional Review

For the four major types of biological macromolecules you just learned about (lipids, proteins, carbohydrates, and nucleic acids), consider how the atoms that make them up fit together to determine the shape of the molecule—as well as how the shape of the molecule is appropriate for its function in the cell.

Multiple-Choice Questions

1. Which of the following functional groups characterizes RNA?
   (A) A PO_3 group, deoxyribose, and uracil
   (B) A PO_3 group, ribose, and uracil
   (C) A PO_3 group, ribose, and thymine
   (D) A PO_2 group, deoxyribose, and uracil
   (E) A PO_2 group, deoxyribose, and thymine

2. Which of the following molecules would contain a polar covalent bond?
   (A) Cl_2
   (B) NaCl
   (C) H_2O
   (D) KBr
   (E) $C_6H_{12}O_6$

3. Which of the following compounds would NOT be water-soluble?
   (A) Potassium chloride
   (B) Fatty acids
   (C) Fructose
   (D) Cellulose
   (E) Hydrogen bromide

4. Three terms associated with the travel of water from the roots up through the vascular tissues of plants are
   (A) adhesion, cohesion, and translocation
   (B) adhesion, cohesion, and transcription
   (C) cohesion, hybridization, and transpiration
   (D) cohesion, adhesion, and transpiration
   (E) transpiration, neutralization, and adhesion

Directions: The group of questions below consists of five lettered headings followed by a list of numbered phrases or sentences. For each numbered phrase or sentence select the one heading that is most closely related to it and fill in the corresponding oval on the answer sheet. Each heading may be used once, more than once, or not at all.

Questions 5–9
   (A) Lipids
   (B) Peptide bonds
   (C) Alpha helix
   (D) Unsaturated fatty acids
   (E) Cellulose

5. Contain one or more double bonds

6. One of the four major classes of biological molecules that are not polymers

7. Linkages between the monomers of proteins

8. A secondary structure of proteins

9. The storage form of carbohydrates in plants

10. The process by which protein conformation is lost or broken down is
    (A) decondensation
    (B) deconstruction
    (C) denaturatrion
    (D) hydrolysis
    (E) hybridization

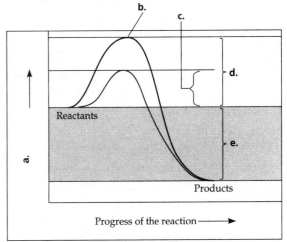

Exergonic Reaction With and Without an Enzyme Catalyst

11. The above graph most accurately depicts the energy changes that take place in which of the following types of reaction?
(A) Hypothermic
(B) Hyperthermic
(C) Endergonic
(D) Exergonic
(E) Free range

12. Which of the following theories or laws states that every energy transfer increases the amount of entropy in the universe?
(A) The free energy law
(B) The first law of thermodynamics
(C) The second law of thermodynamics
(D) Evolutionary theory
(E) The law of increased chaos

13. Catalysts speed up chemical reactions by
(A) decreasing the free energy change of the reaction
(B) increasing the free energy change of the reaction
(C) degrading the competitive inhibitors in a reaction
(D) lowering the activation energy of the reaction
(E) raising the activation energy of the reaction

Directions: The group of questions below consists of five lettered headings followed by a list of numbered phrases or sentences. For each numbered phrase or sentence select the one heading that is most closely related to it and fill in the corresponding oval on the answer sheet. Each heading may be used once, more than once, or not at all.

Questions 14–18
(A) Allosteric inhibition
(B) Feedback inhibition
(C) Competitive inhibitor
(D) Noncompetitive inhibitor
(E) Cooperativity

14. Describes inhibition by an enzyme that is capable of either activating or inhibiting a metabolic pathway

15. A reversible inhibitor that looks similar to the normal substrate and competes for the active site of the enzyme

16. The process by which the binding of the substrate to the enzyme triggers a favorable conformation change, which causes a similar change in all of the proteins' subunits

17. The process by which a metabolic pathway is shut off by the product it produces

18. Binds to the enzyme at a site other than the active site, causing the enzyme to change shape and be unable to bind to the substrate

19. $A + B \rightarrow AB + energy$
Which of the following best characterizes the reaction represented above?
(A) Metabolism
(B) Anabolism
(C) Catabolism
(D) Endergonic reaction
(E) Exergonic reaction

20. When a water molecule loses an H+ in solution, and another water molecule accepts the proton, what is the water molecule that accepts the proton called?
(A) A hydroxide ion
(B) Crystal lattice structure
(C) A solution not in equilibrium
(D) A hydronium ion
(E) A basic solution

Free-Response Question

1. Phospholipids are a critical component of the cell wall. The cell wall is selectively permeable, allowing only certain substances in and out, and in certain amounts.
   ▌ Describe why you think phospholipids are good candidates to form the cell wall, based on their structure and properties.
   ▌ Explain why proteins are an important component of the cell wall, based on their structure and properties.

ANSWERS AND EXPLANATIONS

Multiple-Choice Questions

▌ 1. **(B) is correct.** RNA is made up of a phosphate group, a ribose sugar, and one of the following four nitrogenous bases—cytosine, guanine, uracil, and adenine. The phosphate group of RNA contains a phosphate atom and three atoms of oxygen, not two. DNA is similar to RNA in many ways, but different in two important ones. First, it contains deoxyribose as its sugar, instead of ribose, and it contains the bases cytosine, guanine, thymine, and adenine. DNA does not contain uracil, and RNA does not contain thymine.

▌ 2. **(C) is correct.** The answer is water, H_2O. Polar covalent bonds are those in which valence electrons are shared between atoms, but unequally. (The more electronegative atom will attract the electrons more strongly, and that end of the molecule will have a slight negative charge, whereas the less electronegative atom will attract them less strongly and be slightly positive.) The two atoms involved in the bond must be of differing electronegativities in order to form a polar covalent bond.

▌ 3. **(B) is correct.** Water dissolves polar and ionic compounds, and it doesn't dissolve oils and nonpolar compounds as well. Potassium chloride (KCl) looks a lot like table salt, NaCl. Potassium is positively charged, and chlorine is negatively charged—this is an ionic compound. The same thing applies to hydrogen bromide (HBr), so you can eliminate answer choices *A* and *E*. Sugars dissolve in water (and fructose is a sugar), so you can eliminate *C*. Some hydrophilic substances don't dissolve in water because their molecules are too big. Cellulose (answer choice *D*) is one of these, so technically this is a water soluble, or hydrophilic compound. Now look at choice *B*, and think back to the cell membrane. The phospholipids that make up the cell membrane consist of hydrophobic fatty acid chains, and hydrophilic head groups, which (among other variable components) contain glycerol. The fact that phospholipids have a hydrophilic and a hydrophilic end makes them able to form the cell wall, which is responsible for selectively allowing substances to enter and exit the cell. Fatty acids comprise the hydrophobic end of the phospholipid and are not water soluble.

▌ 4. **(D) is correct.** The three terms you should keep in mind as you think of water traveling up through the xylem of a plant are transpiration (in which water evaporates from the leaves of the plants), cohesion, in which the water molecules stick together due to the hydrogen bonds, and adhesion, whereby the water molecules stick to the sides of the plant cells, helping to pull themselves up through the plant against the force of gravity.

5. **(D) is correct.** Unsaturated fatty acids contain at least one carbon-carbon double bond, whereas saturated fatty acids contain no double bonds.

6. **(A) is correct.** Lipids are the only one of the four major classes of biological molecules that are not polymers. They are grouped together because of their hydrophobicity. Nucleic acids are polymers of nucleotide monomers; proteins are polymers of amino acid monomers; and carbohydrates are polymers of monosaccharide monomers.

7. **(B) is correct.** The linkages between the amino acids of proteins are peptide bonds. Peptide bonds are covalent bonds created in dehydration synthesis reactions. The carboxyl group of one amino acid is joined to the amino group of an adjacent amino acid, resulting in the loss of one molecule of water.

8. **(C) is correct.** One common secondary structure of proteins is the alpha helix; another is the beta pleated sheet. The secondary structure of a protein refers to a section of the polypeptide chain that is repeatedly folded or coiled in a regular pattern. They are the result of regular hydrogen bonding between segments of the polypeptide backbone.

9. **(E) is correct.** Cellulose is the polysaccharide that forms the strong cell walls of plant cells. It is a polymer of glucose. In animals, polysaccharides are stored as glycogen, and chitin is a structural polysaccharide found in arthropods.

10. **(C) is correct.** The process by which a protein loses its overall structure, or conformation, is called denaturation. Denatured proteins are biologically inactive, and denaturation can occur as a result of extreme conditions of pH, temperature, or salt concentration.

11. **(D) is correct.** The shape of the curve in the art shown most closely depicts an exergonic reaction. The potential energy of the products is lower than that of the reactants—meaning that in the course of the reaction, energy is given off. This is characteristic of exergonic reactions. Conversely, in an endergonic reaction, energy is taken in during the course of the reaction.

12. **(C) is correct.** The second law of thermodynamics states that every energy transfer that occurs increases the amount of entropy in the universe. The first law of thermodynamics states that the amount of energy in the universe is constant, and therefore energy can be neither created nor destroyed. Evolutionary theory refers to the myriad changes that have taken place to transform living organisms from the beginning of life on Earth until today.

13. **(D) is correct.** Catalysts speed up chemical reactions by providing an alternate reaction pathway that lowers the activation energy of the reaction. Less energy is required to start the reaction, so it runs more quickly.

14. **(A) is correct.** In allosteric regulation, the enzyme is usually composed of more than one polypeptide chain with more than one allosteric site (remote from the active site), and the enzyme usually oscillates between an inactive conformation and an active one. When an allosteric activator binds to the allosteric site, the protein assumes a stable conformation with a functional active site, and the reaction can proceed. When an allosteric inhibitor binds, this stabilizes the inactive conformation of the protein.

15. **(C) is correct.** Competitive inhibitors bind to the active site of the enzyme through covalent bonds. They are able to bind because they closely resemble the normal substrate. One way to overcome the effects of competitive inhibitors is to increase the amount of substrate so that chances are greater that a substrate molecule (rather than the competitive inhibitor) will bind.

16. **(E) is correct.** In cooperativity, the enzyme in question has more than one subunit with more than one active site, and it is able to bind more than one substrate—so multiple reactions can be taking place at once in the enzyme. The binding of one substrate molecule to the enzyme causes a conformation change that makes the binding of other substrate molecules, at the other active sites, more favorable.

17. **(B) is correct.** In feedback inhibition, the end product of a metabolic pathway switches off the pathway by binding to and inhibiting an enzyme involved somewhere along the pathway.

18. **(D) is correct.** In noncompetitive inhibition, the inhibitor binds to a site other than the active site of the enzyme, and this causes the enzyme to change shape. The change in conformation makes the substrate unable to bind to the active site of the enzyme, and this prevents the reaction from taking place.

19. **(E) is correct.** The reaction shown here is an exergonic reaction. An exergonic reaction is a spontaneous chemical reaction in which there is a net release of free energy. Energy is given off in the course of the reaction shown.

20. **(D) is correct.** In this situation, the water that donated the H^+ would become a hydroxide ion (OH^-), whereas the water that accepted the H^+ would become a hydronium ion, H_3O^+.

Free-Response Question

1. Phospholipids are made up of a negatively charged hydrophilic head (containing a glycerol molecule and a phosphate group) and two hydrophobic fatty acid tails. When in water, the tails are hydrophobic and the heads are hydrophilic, so phospholipids self-assemble into circles, or micelles, in which the hydrophilic heads are all pointing outward while the hydrophobic tails are pointing inward. In creating the cell membrane surface, phospholipids are arranged in a double layer, called a bilayer, in which the hydrophilic heads are in contact with the cell's interior and exterior, while the tails are pointed toward each other, toward the interior of the membrane. The fatty acid chains of phospholipids can contain double bonds, which makes them unsaturated. These double bonds tend to cause bends in the two tails. These kinks in the tails of phospholipids contribute to the fluidity of the cell membrane because the phospholipids aren't packed together tightly. The fluidity of the cell membrane is very important in its function; the less fluid the membrane is, the more impermeable it is. There is an optimum permeability for the cell membrane, at which all of the substances necessary for metabolism can pass into and out of the cell.

   A significant reason why phospholipids are the perfect major component of the cell wall is their dual nature of being hydrophilic and hydrophobic. This nature makes the cell membrane selectively permeable; it enables hydrophobic molecules such as hydrocarbons, carbon dioxide, and oxygen, to dissolve in the bilayer and easily cross the membrane. However, ions and polar molecules (including water, glucose, and other sugars) can't get through, because of the hydrophobic interior of the membrane. Fortunately, the cell can selectively take up substances from the environment, through protein channels and transporters.

Proteins are perfect cell membrane transporters because they are able to act as channels; substances that bind to them can help alter their conformation to permit the passage of molecules through them, and into the cell interior.

There are many different ways by which proteins can permit the passage of ionic and polar molecules through the lipid bilayer. Proteins associated with the membrane are either integral proteins, which actually penetrate the lipid bilayer (ones that completely go through the bilayer are called transmembrane proteins), or they are "peripheral proteins" that are just associated with the outside of the membrane. Transmembrane proteins can form hydrophilic channels that permit the passage of certain hydrophilic substances that otherwise would not be able to get across. Other functions that membrane proteins serve are to attach the cell to the extracellular matrix, to stabilize it, and to function in cell-cell recognition. Membrane proteins are also important in cell-cell signaling; some have enzyme function and carry out important metabolic reactions, and they aid in joining adjacent cells.

This response shows thorough knowledge of the processes of the structure of phospholipids, cell membrane structure and components, and movement across membranes. A strong response to this item requires an understanding of topics from Units 1 and 2 of the textbook. Note that the response includes the following key terms in context, showing the writer's knowledge of their meanings and relatedness:

phospholipids	*double bonds*	*transporters*
hydrophilic head group	*unsaturated*	*conformation*
glycerol	*permeability*	*integral proteins*
phosphate	*metabolism*	*transmembrane proteins*
hydrophobic	*hydrocarbons*	*extracellular matrix*
lipid bilayer	*protein channels*	*cell-cell signaling*

CHAPTER 2

The Cell

A Tour of the Cell: How We Study Cells

▌ **Light microscopes** (LSs) are used to observe most plant and animal cells, bacteria, and some organelles like mitochondria and nuclei, although most cell organelles are too small to be seen with a light microscope. With these instruments, scientists can observe things from about 1mm to about 1mm in size.

▌ Electron microscopes are used to study objects from about 0.1 nm to 100 mm in size. They function by focusing a beam of electrons either through the specimen or onto its surface. There are two kinds of EMs: **transmission electron microscopes** (TEMs) and **scanning electron microscopes (SEMs)**.

A Panoramic View of the Cell

The best way to remember the main facts about **prokaryotes** and **eukaryotes** is to study a table of their major characteristics:

Characteristics	Prokaryotic Cells	Eukaryotic Cells
Plasma membrane	yes	yes
Cytosol with organelles	yes	yes
Ribosomes	yes	yes
Nucleus	no	yes
Size	1µm – 10 µm	10 µm – 100 µm
Internal membranes	no	yes

Prokaryotic cells include bacteria and archaebacteria, whereas eukaryotic cells are animal and plant cells. Some details to remember about **prokaryotes** include:

■ no membrane-bound nucleus—instead chromosomes grouped together in region called the nucleoid
■ no membrane-bound organelles
■ smaller than eukaryotes
■ consist of bacteria and archaebacteria

Some details to remember about **eukaryotic** cells include:

■ membrane-bound nucleus, which contains cell's chromosomes
■ membrane-bound organelles in cytoplasm
■ much larger than prokaryotes
■ eukaryotes consist of protists, fungi, plants, and animal cells

The Nucleus and Ribosomes, the Endomembrane System, Other Membranous Organelles, and the Cytoskeleton

You should be familiar with the structures that follow this illustration.

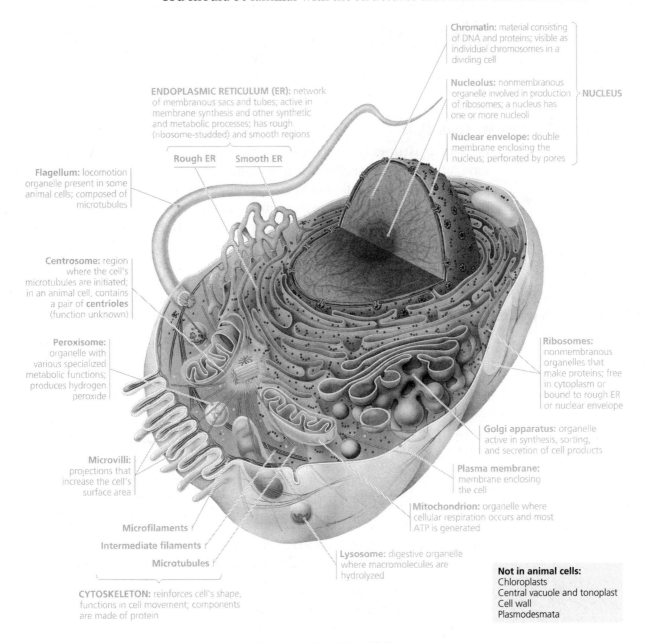

Chromatin: material consisting of DNA and proteins; visible as individual chromosomes in a dividing cell

Nucleolus: nonmembranous organelle involved in production of ribosomes; a nucleus has one or more nucleoli

Nuclear envelope: double membrane enclosing the nucleus; perforated by pores

NUCLEUS

ENDOPLASMIC RETICULUM (ER): network of membranous sacs and tubes; active in membrane synthesis and other synthetic and metabolic processes; has rough (ribosome-studded) and smooth regions

Rough ER   Smooth ER

Flagellum: locomotion organelle present in some animal cells; composed of microtubules

Centrosome: region where the cell's microtubules are initiated; in an animal cell, contains a pair of **centrioles** (function unknown)

Peroxisome: organelle with various specialized metabolic functions; produces hydrogen peroxide

Ribosomes: nonmembranous organelles that make proteins; free in cytoplasm or bound to rough ER or nuclear envelope

Golgi apparatus: organelle active in synthesis, sorting, and secretion of cell products

Plasma membrane: membrane enclosing the cell

Microvilli: projections that increase the cell's surface area

Mitochondrion: organelle where cellular respiration occurs and most ATP is generated

Microfilaments

Intermediate filaments

Microtubules

Lysosome: digestive organelle where macromolecules are hydrolyzed

CYTOSKELETON: reinforces cell's shape, functions in cell movement; components are made of protein

Not in animal cells:
Chloroplasts
Central vacuole and tonoplast
Cell wall
Plasmodesmata

Overview of an Animal Cell

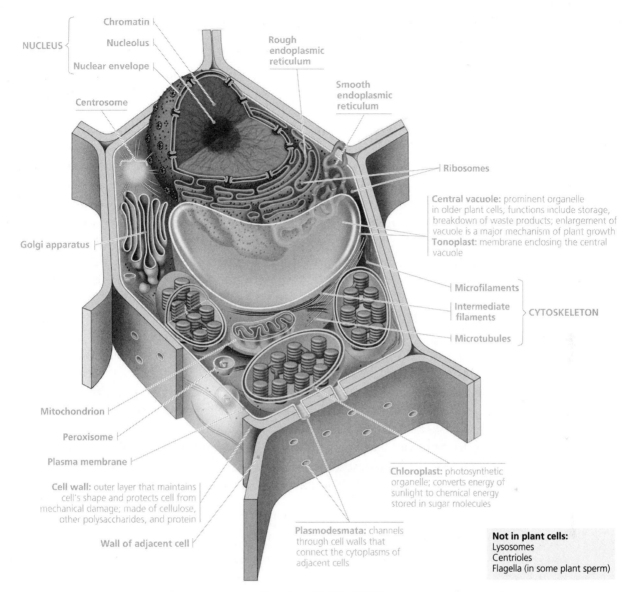

Overview of a Plant Cell

Cell structures found in both plant and animal cells

▌ **plasma membrane**—creating the boundary for a cell, selectively permits the passage of materials into and out of the cell, made up of phospholipids, proteins, and associated carbohydrates

▌ **nucleus**—contains most of the cell's chromosomes; the most noticeable organelle in the cell because of its relative size; surrounded by a double membrane, mRNA is transcribed here and then sent out to the cytoplasm; considered the control center of the cell

▌ **chromatin**—the complex of DNA and protein housed in the nucleus; as cell gets ready for cell division, chromatin condenses into **chromosomes**; each species has unique number of chromosomes (humans with 46 chromosomes in their somatic cells)

- **nucleolus**—associated with chromatin, the nucleolus exists in a non-dividing nucleus; Ribosomal RNA (rRNA) is produced here and is not enclosed by a membrane
- **ribosomes**—made up of rRNA and protein, ribosomes are sites of protein synthesis in the cell; consist of two subunits and not enclosed by a membrane; two types of ribosomes—those free-floating in cytosol, and those bound to endoplasmic reticulum; both produce proteins
- **endoplasmic reticulum** (ER)—makes up more than half the total membrane structure in many cells; a network of membranes and sacs called cisternae; its internal component called the cisternal space; two types of ER—rough and smooth ER; **smooth ER** involved in synthesis of lipids and metabolism of carbohydrates; **rough ER** so called because of its associated ribosomes; ribosomes associated with ER synthesize important proteins secreted by the cell; as these proteins are created on the ER-bound ribosomes, the polypeptide chain travels through ER membrane into cisternal space, where they are kept separate from proteins created at free-floating ribosomes, which will not be secreted; rough ER proteins are then excreted from cell through transport vesicles
- **Golgi apparatus**—when transport vesicles from the rough ER leave the ER, they next travel to the Golgi apparatus, where their contents are modified, stored, then sent on their way; Golgi apparatus consists of flattened sacs of membranes, again called cisternae; Golgi stacks have polarity—one face is called the *cis* face and the other is called the trans face; *cis* face receives vesicles and *trans* face ships vesicles out
- **mitochondria**—organelles in which cell respiration takes place; in cell respiration, ATP is created, so mitochondria are often referred to as the powerhouses of the cell; enclosed by a double membrane; inner membrane has infolds called **cristae**
- **peroxisomes**—single-membrane bound compartments in cell responsible for various metabolic functions, like breakdown of purines; produce hydrogen peroxide and also break down hydrogen peroxide in plant cells
- **cytoskeleton**—a network of fibers that runs through the entire cytoplasm; responsible for organizing cell structures and activities; involved in cell motility, including actual locomotion of cell, and movement of structures within cell; three types of fibers make up cytoskeleton—**microtubules** (which are made up of the protein **tubulin**), **microfilaments**, and **intermediate filaments**
- **cilia**—appendages used for cell locomotion, protruding from cell surface in large numbers

Cell structures found only in animal cells

- **lysosomes**—membrane-bound sacs of hydrolytic enzymes that can digest large molecules, such as proteins, polysaccharides, fats, and nucleic acids; have a low internal pH; can break down macromolecules for organic monomers to be excreted back into cytosol and recycled by cell; as a cell ages, the lysosomal membrane can break, thus digesting its cellular contents and destroying the cell—hence the nickname "the suicide bag"

- **centrosomes**—a region located near the nucleus in animal cells, from which microtubules grow
- **centrioles**—located within centrosome of animal cells; replicate before cell division
- **flagella**—appendages used for cell locomotion, protruding from cell surface; most cells with flagella possess only one, but a few possess more than one
- **extracellular matrix** of animal cells—situated just external to plasma membrane; composed of glycoproteins secreted by cell (most prominent of which is collagen)
- **tight junctions**—sections of animal cell membrane where two neighboring cells are fused
- **desmosomes**—fasten adjacent animal cells together; made up of intermediate filaments
- **gap junctions**—provide channels between adjacent animal cells through which ions, sugars, and other small molecules can pass

Cell structures found only in plant cells

- **central vacuoles**—membrane-bound organelles whose functions include storage and breakdown of some waste products; in plants, a vacuole can make up about 80% of cell
- **chloroplasts**—found in both plant and algae cells; sites of photosynthesis; here solar energy is converted to chemical energy; inside chloroplasts are flattened membranes called thylakoids; stacked thylakoids are called grana
- **cell wall** of a plant—protects the plant and helps maintain its shape
- **plasmodesmata**—channels that perforate adjacent plant cell walls and allow cytosol to pass through from cell to cell

Membrane Structure and Function

Membranes are of utmost important to the cell as a whole, and to many of the organelles contained in the cell, because they act as selective barriers to let in only the substances that each cell or specific organelle needs to function properly.

- Membranes are primarily made up of phospholipids and proteins (though carbohydrates are crucial to membranes, too) held together by weak interactions that cause the membrane to be fluid. In the **fluid mosaic model** of the cell membrane, the membrane is fluid, and proteins are embedded in or associated with the phospholipid bilayer.
- There are both integral proteins and peripheral proteins in the cell membrane. **Integral proteins** are those that are completely embedded in the membrane, some of which are transmembrane proteins that span the membrane completely. **Peripheral proteins** are loosely bound to the membrane's surface.
- **Carbohydrates** on the membrane are crucial in cell-cell recognition (which is necessary for proper immune function) and in developing organisms (for tissue differentiation). Cell surface carbohydrates—many of which are oligosaccharides—vary from species to species and are the reason that blood transfusions must be type-specific.

▌ **Hydrocarbons**, **carbon dioxide**, and **oxygen** are hydrophobic substances that can pass easily across the cell membrane. **Ions** and **polar molecules** cannot pass easily across the membrane. The former substances move across the membrane through **passive diffusion**. In passive diffusion, the substance will travel from where it is more concentrated to where it is less concentrated, diffusing down its **concentration gradient**. This type of diffusion requires that no work be done, and it relies only on the thermal motion energy intrinsic to the molecule in question. The term "passive diffusion" is used because the cell expends no energy in moving the substances.

▌ The word for the passive transport of water is **osmosis**. In osmosis, water flows from the **hypotonic solution** (the solution with lower solute concentration) to a **hypertonic solution** (one with higher solute concentration).

▌ Hydrophilic substances get across the membrane through **transport proteins**. Transport membranes work in two ways:

1. They provide a hydrophilic channel through which the molecules in question can pass.
2. They bind loosely to the molecule in question and carry them through the membrane.

▌ The process by which ions and hydrophilic substances diffuse across the cell membrane with the help of transport proteins is called **facilitated diffusion**. Transport proteins are specific for the substances they transport.

▌ In **active transport**, substances can be moved against their concentration gradient. Not surprisingly, in this type of transport, the cell must expend energy. This type of transport is crucial for cells to be able to maintain sufficient quantities of substances that are relatively rare in their environment.

▌ Specific transmembrane proteins are responsible for active transport, and **ATP** supplies the energy for this type of transport. ATP transfers one of its phosphates to the transport protein, which might be responsible for making the protein change its shape to allow for the passage of the substance.

▌ Ions have both a chemical and a voltage gradient across the membrane, and this causes an **electrochemical gradient**. The inside of the cell is slightly more negative than the outside, so that membrane potential favors the movement of cations (positively charged ions) into the cell.

▌ The sodium-potassium pump works by exchanging sodium (Na^+) for (K^+) across the cell membrane; it exchanges three Na ions for two K ions, so for each round of the pump, there is a net transfer of one positive charge from the cell interior to the exterior. This type of pump, called an **electrogenic pump**, generates voltage across the membrane.

▌ In **cotransport**, an ATP pump that transports a specific solute indirectly drives the active transport of other substances. In this process, the substance that was initially pumped across the membrane will do work, providing energy for the transport of another substance against its concentration gradient, as it leaks back across the membrane with its concentration gradient.

▌ Large molecules are moved across the cell membrane through exocytosis and endocytosis. In **exocytosis**, vesicles from the cell's interior fuse with the cell membrane expelling their contents to the exterior. In **endocytosis**, the cell forms new vesicles from the plasma membrane; this is basically the reverse of exocytosis, and this process allows the cell to take in macromolecules. There are three types of endocytosis:

1. **Phagocytosis** occurs when the cell wraps pseudopodia around the substance and packages it within a large vesicle formed by the membrane.
2. In **pincocytosis**, the cell takes in small droplets of extracellular fluid in small vesicles. Pinocytosis is not specific—no target molecules are taken in, in this process.
3. **Receptor-mediated endocytosis** is a very specific process. Certain substances (ligands) bind to specific receptors on the cell's surface (these receptors are usually clustered in **coated pits**), and this causes a vesicle to form around the substance and then to pinch off into the cytoplasm.

Cellular Respiration: Harvesting Chemical Energy

Coupled reactions (otherwise known as oxidation-reduction reactions), fermentation, cell respiration, and photosynthesis are covered in one of the most technically challenging sections of your textbook—since much of the material is based on chemistry. Here we will focus on the major steps of each of the processes, as well as the end results. Questions on the AP Biology Exam are likely to focus on the net results of photosynthesis and respiration—not on the exact reactions that create the products.

Principles of Energy Harvest

▌ **Energy** in a cell is stored in the chemical bonds between atoms, and this energy is released when the bonds are broken. Some of the energy that is released does work for the cell, and the rest is given off as heat.

▌ **Catabolism** is the process by which molecules are broken down and their energy is released. Two types of catabolism are:

1. **fermentation**—the breakdown of sugars that occurs in the absence of oxygen
2. **cellular respiration**—the breakdown of sugars that occurs in the presence of oxygen
   Carbohydrates, fats, and proteins can all be broken down to release energy in cell respiration. However, glucose is the primary nutrient fuel molecule that is used in cellular respiration to release energy. The standard way of representing the process of cell respiration shows glucose being broken down in the following reaction:

$$C_6H_{12}O_6 + 6O_6 \rightarrow 6CO_2 + 6H_2O + \text{Energy (in the form of ATP and heat)}$$

As you can see from the reaction, the breakdown of glucose is exergonic—that is, energy is given off in this reaction, in the form of ATP. This ATP is used to power all cellular activities.

- **ATP**, or **adenine triphosphate**, is released as energy when enzymes in the cell transfer one of its three phosphate groups to another molecule. ATP can be recycled. The dephosphorylated ATP is ADP (adenine diphosphate), and the ADP can be re-phosphorylated by the cell to form ATP again.
- To fully understand how energy is exchanged in these molecular reactions in the cell, we need to look at exactly how energy moves. As you learned very early in your biology course, electrons are exchanged in the course of chemical reactions. When electrons are transferred from one reactant to another, the reaction is called a **oxidation-reduction reaction**, or **redox reaction**.
- The loss of an electron from a substance is called **oxidation**; the substance that gives up the electron is said to be **oxidized**. The gain of an electron by a substance is called **reduction**; the substance that accepts the electron is said to be **reduced**.

The Process of Cellular Respiration

There are three main stages of respiration: glycolysis, the Krebs cycle, and the electron transport chain and oxidative phosphorylation.

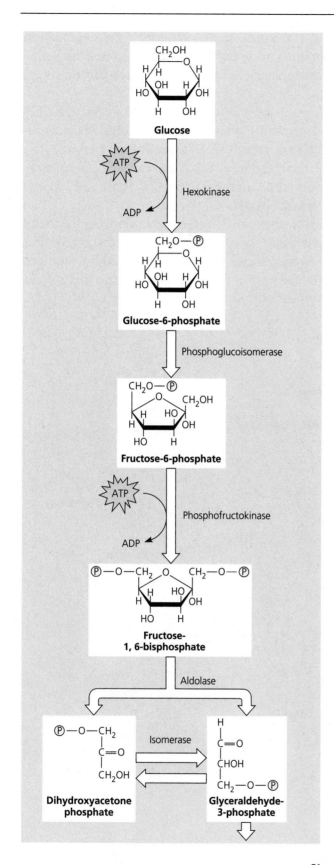

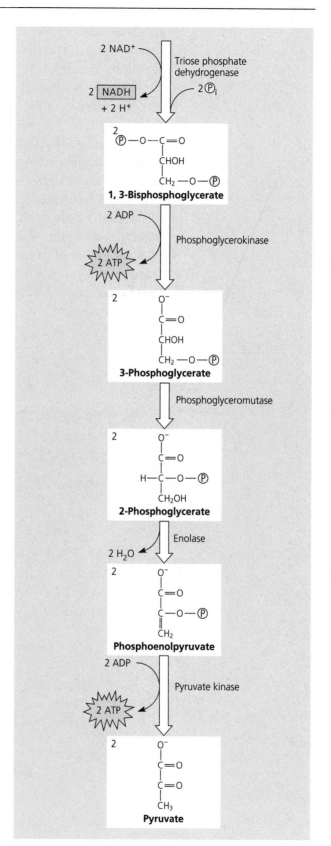

Glycolysis

- In **glycolysis** (which occurs in the cytosol), glucose begins the degradation process when it is broken down into two pyruvate molecules. As you can see, in glycolysis, the six-carbon glucose molecule is split into two three-carbon sugars through a long series of steps.

- In the course of glycolysis, there is an ATP-consuming phase and an ATP-producing phase. In the ATP-consuming phase, two ATP molecules are consumed, but later 4 ATP molecules are produced. So there is a net gain of 2 ATP in glycolyis, plus 2 NADH are also produced, and these can be used to make more ATP in oxidative phosphorylation, in the presence of oxygen. So, the **net results of glycolysis** are **2 ATP** and **2 NADH**.

- In the **Krebs cycle** (which occurs in the matrix of the mitochondria), the job of breaking down glucose is completed and the final product is CO_2, carbon dioxide.

- In the presence of oxygen, the two **pyruvate** molecules produced by glycolysis travels to the mitochondrion to take part in the Krebs cycle. In the Krebs cycle, the chemical energy that remains in the pyruvate molecules is released.

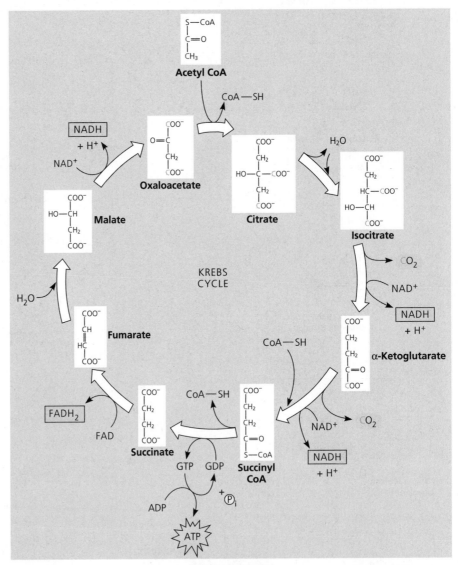

The Krebs Cycle

▌ The Krebs cycle has 8 steps, and like glycolysis, each of the steps is catalyzed by a different enzyme. Each cycling of the Krebs cycle requires the input of a 2-carbon acetyl-CoA molecule, and two carbons are released in the course of the cycle, as CO_2. The **net results of the Krebs cycle** are:

> **4 CO_2**
> **2 ATP**
> **6 NADH**
> **2 $FADH_2$**

Since glycolysis results in the production of *two* pyruvate molecules (which are converted to acetyl-CoA) and each turn of the Krebs cycle requires the input of one acetyl-CoA. The Krebs cycle must make two turns before you get the products from the breakdown of the two pyruvate molecules from the Krebs cycle.

▌ So far, glycolysis and the Krebs cycle have produced a total of 4 molecules of ATP for each glucose molecule that was degraded. The rest of the ATP produced from the breakdown of glucose is produced in the next steps of cell respiration, through the electron transport chain and oxidative phosphorylation.

▌ The NADH and $FADH_2$ that were created during the Krebs cycle are used in the course of the electron transport system and oxidative phosphorylation to create relatively large amounts of ATP.

▌ The **electron transport chain** produces energy that drives the synthesis of ATP in **oxidative phosphorylation**. The electron transport chain consists of molecules (mostly proteins) that are embedded in the inner mitochondrial membrane. There are millions of such electron transport chains in the inner mitochondrial membrane.

▌ Sitting atop these proteins embedded in the membrane are associated molecules that are alternately reduced and oxidized as they accept and donate electrons.

▌ The initial electron acceptor in the electron transport chain is a flavoprotein called flavin mononucletide or FMN, and it accepts an electron from NADH. The electron is passed down a series of molecules to oxygen, which is the final electron acceptor. Then it is combined with a couple of hydrogen atoms to form a molecule of water.

▌ $FADH_2$ and NADH both donate electrons to the chain.

▌ The electron transport chain doesn't make any ATP itself. Instead these reactions are coupled to others to produce ATP in a process called chemiosmosis.

▌ The **net result** of the electron transport chain:

1. to move free energy down a series of steps from $FADH_2$ and NADH to oxygen
2. to provide a source of energy for the creation of ATP through chemiosmosis

Oxidative Phosphorylation and Chemiosmosis

▌ Also embedded in the mitochondrial inner membrane are protein complexes called **ATP synthetases**. ATP synthetases phosphorytetes ATP out of ADP plus inorganic phosphate. ATP synthetases use the energy from a proton gradient created by the electron transport system to power the synthesis of ATP.

▌ The flow of electrons in the electron transport chain is exergonic, and the energy given off is used to pump H^+ ions across the membrane against their concentration gradient. The H^+ ions flow back across the membrane into the mitochondrial matrix with their concentration gradient, and since the H^+ ions can only flow back through the ATP synthetases (they are the only regions in this membrane permeable to H^+ ions), their flow drives the oxidative phosphorylation of ADP to ATP. This is how the H^+ gradient created by the electron transport chain is coupled to ATP synthesis, in **chemiosmosis**.

Related Metabolic Processes

In the absence of oxygen (remember that oxygen is the final electron acceptor in the electron transport chain, and without it this chain will not function), the cell goes through a process called fermentation.

▌ Fermentation takes place under **anaerobic** conditions—that is, when there is no oxygen present. Cell respiration takes place under **aerobic** conditions (when oxygen is present). Fermentation consists of glycolysis and reactions that regenerate NAD^+ (so that it can be reused during glycolysis).

▌ The two common types of fermentation are alcohol fermentation and lactic acid fermentation.

  ▪ In **alcohol fermentation**, pyruvate is converted to ethanol, releasing CO_2 and oxidizing NADH in the process to create more NAD^+.
  ▪ In **lactic acid fermentation**, pyruvate is reduced by NADH (and NAD^+ is created in the process), and lactate is formed as a waste product.

Photosynthesis: Photosynthesis in Nature

In plant cells, chloroplasts capture light energy from the sun and convert it to chemical energy that can be stored in sugars and other organic compounds. This process is called **photosynthesis**. Here is some information about the sites of photosynthesis, chloroplasts.

▌ **Chloroplasts** are plant cell organelles that are mostly located in the cells that make up the **mesophyll** tissue, which is part of the plant's leaf. The exterior of the lower epidermis of the leaf cell contains many tiny pores called **stomata**, through which carbon dioxide can enter and oxygen can exit the leaf.

▌ Chloroplasts have an outer membrane and an inner membrane. Inside the inner membrane is the **stroma**, which is a dense fluid-filled area. Within the stroma is a vast network of interconnected **thylakoid membranes**, inside which lies the **thylakoid space**.

▌ **Chlorophyll** is located in the thylakoid membranes and is the light-absorbing pigment that drives photosynthesis and gives plants their green color.

The Pathways of Photosynthesis

The overall reaction of photosynthesis looks like this:

$$6CO_2 + 12H_2O + \text{Light energy} \rightarrow C_6H_{12}O_6 + 6O_2 + 6H_2O$$

Basically, this means that plants can produce organic compounds and oxygen using light energy, carbon dioxide, and water. The two main parts of photosynthesis are the light reactions and the Calvin cycle.

▌ In the **light reactions** of photosynthesis, solar energy is converted to chemical energy. Light is absorbed by chlorophyll and drives the transfer of electrons from water to $NADP^+$ (to create NADPH), which stores them. Water is split during the course of these reactions, and O_2 is given off. The light reactions also produce ATP from ADP in photophosphorylation. The net products of the light reactions are **NADPH** (which stores electrons), **ATP**, and **oxygen**.

▌ In the **Calvin cycle**, CO_2 from the air is incorporated into organic molecules in **carbon fixation.** The fixed carbon is then used to make carbohydrates. NADPH is used to power carbon fixation. The Calvin cycle also uses ATP in the course of its reactions.

The Light Reactions

▌ Light is electromagnetic energy, and it behaves as though it is made up of discrete particles, called **photons**—each of which has a fixed quantity of energy.

▌ Substances that absorb light are called **pigments**, and different pigments absorb light of different wavelengths. Chlorophyll is a pigment that absorbs not only red and blue but also green. This is why we see summer leaves as green.

▌ When chlorophyll absorbs light energy in the form of photons, one of the molecule's electrons is raised to an orbital of higher potential energy. The chlorophyll is then said to be in an "excited" state.

▌ Photons of light are absorbed by certain groups of pigment molecules in the thylakoid membrane of chloroplasts. These groups are called **photosystems**. Photosystems have an antenna complex made up of chlorophyll molecules and caretenoid molecules (accessory pigments in the thylakoid membrane); this allows them to gather light effectively.

▌ There are two photosystems in the thylakoid membrane that are important to photosynthesis—**photosystem I (PSI)** and **photosystem II (PSII).** Each of these photosystems has a **reaction center** (the site of the first light-driven chemical reaction of photosynthesis).

▌ Here are the major steps of the light reactions of photosynthesis.

1. Photosystem II absorbs light in the 680-nanometer wavelength range. An electron in the reaction center chlorophyll (called P680) becomes excited and then captured by a primary electron acceptor. The reaction center chlorophyll is oxidized and needs an electron.

2. An enzyme supplies the missing electron taken from **photolysis** water (the splitting of water) to P680; water is split in the process, and a free oxygen is created—this oxygen combines with another oxygen to form O_2.

3. The original excited electron passes from the primary electron acceptor of photosystem II to photosystem I through an electron transport chain.

4. The energy from the transfer of electrons down the electron transport chain is used to phosphorylate ADP to ATP in the thylakoid membrane, in a process called **noncyclic photophosphorylation**. This is the same process as in chemiosmosis. Later this ATP will be used as energy in the formation of carbohydrates, in the dark reactions, or the Calvin cycle.

5. The electrons that get to the end of the electron transport chain are donated to the chlorophyll in P700 of photosystem I. (This need for an electron by PSI is created when light energy excites an electron in P700, and that electron is taken up by the primary acceptor of photosystem I).

6. The primary electron acceptor of photosystem I passes along the excited electrons along to another electron transport chain, which transmits them to ferredoxin, and then finally to $NADP^+$, which is reduced to **NADPH**, the second of the two important light-reaction products.

▌ An alternative to noncyclic electron flow is **cyclic electron flow**. While noncyclic electron flow produces nearly equal quantities of ATP and NADPH, the Calvin cycle reactions use more ATP than NADPH. In cyclic electron flow, photosystem II is bypassed, and the electrons from ferredoxin cycle back to a portion of the electron transport chain of PSII and its cytochromes and then to P700. Neither NADPH nor oxygen is produced, but ATP is still a product. This process is called **cyclic photophosphorylation**. Cyclic photophosphorylation can occur in some photosynthetic bacteria.

The Calvin Cycle

▌ In the course of the **Calvin cycle**, CO_2 is converted to a carbohydrate called glyceraldehyde-3-phosphate (G3P), and ATP and NADPH are both consumed. But in order to make one molecule of G3P, the cycle must go through three rotations and fix three molecules of CO_2. Here is an outline of what the cycle looks like:

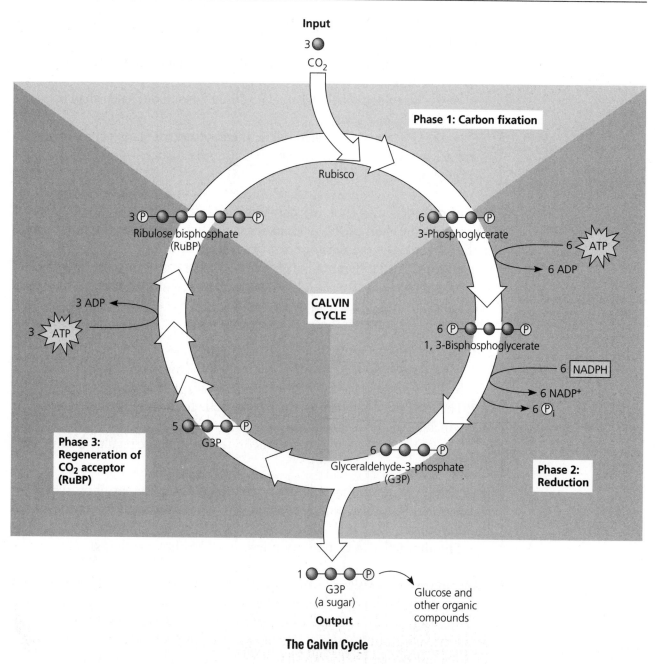

Input

3 ●
CO_2

Phase 1: Carbon fixation

Rubisco

3 Ⓟ ●●●●● Ⓟ
Ribulose bisphosphate
(RuBP)

6 ●●● Ⓟ
3-Phosphoglycerate

6 ✦ ATP
6 ADP

**CALVIN
CYCLE**

3 ADP

3 ✦ ATP

6 Ⓟ ●●● Ⓟ
1, 3-Bisphosphoglycerate

6 NADPH
6 NADP⁺
6 Ⓟᵢ

5 ●●● Ⓟ
G3P

6 ●●● Ⓟ
Glyceraldehyde-3-phosphate
(G3P)

**Phase 3:
Regeneration of
CO_2 acceptor
(RuBP)**

**Phase 2:
Reduction**

1 ●●● Ⓟ
G3P
(a sugar)

Glucose and
other organic
compounds

Output

The Calvin Cycle

These are the steps of the Calvin cycle:

1. The three CO_2 molecules are attached to three **ribulose bisphosphate molecules (RuBPs)**; these reactions are catalyzed by **rubisco** to produce an unstable product that immediately splits into two three-carbon compounds called **3-phosphoglycerate**.
2. The 3-phosphoglycerate molecules are phosphorylated to become **1, 3-disphosphoglycerate**.
3. Next, NADPHs reduce the 1, 3-disphosphoglycerates to create **glyceraldehyde-3-phosphates** (G3P).
4. Finally, RuBP is regenerated as the 5 G3Ps are reworked into 3 of the starting molecules, with the expenditure of 3 ATP molecules.

The results of the Calvin cycle (which produces one G3P molecule for each trip through the cycle) are that:

- 9 molecules of ATP are consumed (to be replenished by the light reactions);
- 6 molecules of NADPH are consumed (also to be replenished by the light reactions);
- the G3P that was created is later metabolized into larger carbohydrates.

There are two other ways in which plants can fix carbon—C_4 photosynthesis and CAM photosynthesis.

▌ Lots of plants living in hot, dry climates use **C_4 fixation** instead of C_3 fixation (the standard Calvin cycle). In C_4 fixation, the first carbon compound formed in the Calvin cycle contains four carbons instead of three.

▌ C_4 plants have two different kinds of photosynthetic cells: **bundle-sheath cells** and mesophyll cells. The bundle-sheath cells are grouped around the leaf's veins, and the mesophyll cells are dispersed elsewhere around the leaf.

▌ The steps of C_4 photosynthesis are as follows:

1. CO_2 is added to **phosphoenolpyruvate** (PEP) to form the four-carbon compound **oxaloacetate** or oxaloacetic acid This enzyme is catalyzed by **PEP carboxylase**, and this process is very quick and efficient.
2. The mesophyll cells export the oxaloacetate to the bundle sheath cells, which break the oxaloacetate back down into CO_2.
3. The CO_2 is converted into carbohydrates through the regular Calvin cycle.

Basically C_4 photosynthesis is just a way to speed up regular photosynthesis, since PEP carboxylase works much faster than rubisco, the enzyme of C_3 photosynthesis.

The alternative to C_3 photosynthesis is CAM photosynthesis.

▌ **CAM photosynthesis** is also an adaptation to hot dry climates. The plants that participate in this kind of photosynthesis open their stomata at night and close them during the day, so that they experience minimal water loss during the day when the sun is out. (Non-CAM plants do the opposite of this.) Because their stomata are closed during the day, CO_2 can't get into the leaves during the day—so they take it up at night.

▌ CAM plants take CO_2 into their leaves at night, convert it into various organic compounds, and put it into temporary storage in their vacuoles. When morning comes and their stomata close, these plants release the stored CO_2 so that they can use light energy to perform the normal reactions of photosynthesis.

▌ In both C_4 and CAM photosynthesis, CO_2 is first transformed into an organic intermediate before it enters the Calvin cycle. All of the processes—C_3, C_4 and CAM photosynthesis—use the Calvin cycle; they just have different methods for getting there.

The Cell Cycle: The Key Roles of Cell Division

The overall cell cycle, the process of mitosis, and cell cycle regulation are topics that appear often in AP Biology exams, and you will be expected to know this material inside and out.

▌ Before the cell can divide, the cell's genome (its complete complement of DNA, in the form of chromosomes) must be copied. All eukaryotic organisms have a characteristic number of chromosomes in their cell nuclei. In human somatic cells (meaning non-sex or gamete cells), this number is 46 chromosomes, which is the diploid chromosome number. Human gametes—sperm and egg cells—which are haploid, have 23 chromosomes. Mitosis is the process by which somatic cells divide and create daughter cells that contain the same diploid chromosome number as the organism.

▌ When the chromosomes are replicated, just prior to mitosis, each duplicated chromosome consists of two sister **chromatids** attached by a **centromere**.

▌ **Mitosis** is the division of the cell's nucleus. It is followed by **cytokinesis**, which is the division of the cell's cytoplasm. Human somatic cells start with 46 homologous chromosomes, 23 of which were inherited from each parent. The chromosomes are doubled, but then mitosis reduces the chromosome number back to its regular 46 number again. Thus, somatic cell reproduction always results in offspring cells with the same diploid chromosome number as their parents.

▌ Conversely, gametes are produced by a process called meiosis, which occurs only in the ovaries and testes of humans. In **meiosis**, daughter cells have half as many chromosomes (23) as their parents have.

The Mitotic Cell Cycle and Mitosis

Let's now outline the entire cell cycle, including the process of mitosis. The cell cycle consists of:

- the **mitotic phase** (10% of the total cycle, consisting of mitosis and cytokinesis);
- **interphase** (90% of the cell cycle, consisting of G_1 phase, S phase, and G_2 phase).

During interphase, the cell prepares for cell division by growing, duplicating cell organelles, replicating histones and other proteins associated with DNA, and replicating its DNA.

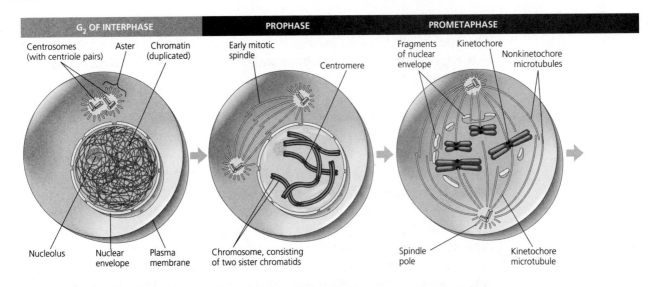

| G₂ OF INTERPHASE | PROPHASE | PROMETAPHASE |

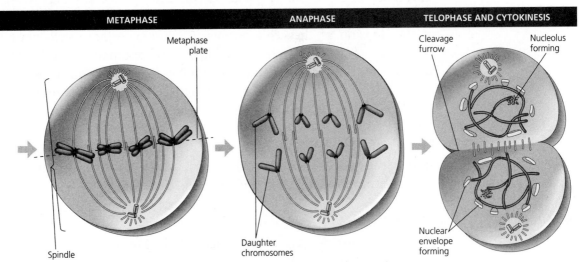

| METAPHASE | ANAPHASE | TELOPHASE AND CYTOKINESIS |

The Stages of Mitotic Cell Division in an Animal Cell

Mitosis can be broken down into five phases, not including cytokinesis. It might be helpful to think of a good mnemonic for these steps.

▎**Prophase:** In this phase, the chromatin becomes more tightly coiled into discrete chromosomes, the nucleoli disappear, and the mitotic spindle (consisting of microtubules extending from the two centrosomes) begins to form in the cytoplasm.

▎**Prometaphase:** The nuclear envelope begins to fragment so that the microtubules can begin to attach to the chromosomes, which have further condensed. Each of the two chromatids from the chromosome pair now has a kinetochore at its centromere region, to which the microtubules will begin to attach. Also, the nucleoli have disappeared.

▎**Metaphase:** In this phase, the centrioles have now migrated to opposite poles in the cell, and the chromosomes line up on the metaphase plate at the equator of the cell. All of the kinetochores have attached microtubules, and all the microtubules together are called the spindle at this point.

- **Anaphase:** In this phase, the sister chromatids begin to separate, pulled apart by the retracting microtubules. The cell also elongates at this time, as the poles elongate. By the end of anaphase, the opposite ends of the cell both contain complete and equal sets of chromosomes.
- **Telophase:** At this point, nuclear envelopes re-form around the sets of chromosomes located at opposite ends of the cell. The chromatin fiber of the chromosomes becomes less condensed, and cytokinesis begins, during which the cytoplasm of the cell is divided. In animal cells, a **cleavage furrow** is created that eventually pinches off to form two cells; in plant cells, a cell plate is created that eventually creates two daughter cells.

You will want to be familiar with how the microtubules reel in the chromosomes. According to research, the kinetochores are equipped with protein motors that "walk" along the microtubule, breaking off tubulin subunits as they go, thereby shortening the microtubules and effectively reeling in the chromosomes. This process is ATP-dependent.

Regulation of the Cell Cycle

The steps of the cycle are known to be controlled by a cell cycle control system. This control system moves the cell through its stages by a series of checkpoints, at which signals tell the cell either to continue or to stop.

- The major cell cycle checkpoints include the **G_1 phase checkpoint, G_2 phase checkpoint,** and **M phase checkpoint.** The **G_1 phase checkpoint** is thought of as the most important checkpoint. If the cell gets the go ahead signal at this checkpoint, it will complete the whole cell cycle and divide. If it does not receive the go-ahead signal, it enters a nondividing phase called G_0 phase.
- **Kinases** are the proteins that control the cell cycle. They exist in the cells at all times but are active only when they are connected to cyclin proteins. Thus they are called **cyclin-dependant kinases (CDK).** As you might guess, the levels of cyclin and CDK activity rise and fall simultaneously in the cell cycle.
- **MPF,** or **maturation-promoting factor,** is a CDK. MPF triggers the cell to pass the G_2 checkpoint into M phase.

For Additional Review

Compare the process of meiosis with the process of mitosis, including in your comparison a study of the change in chromosome number through the cell, the purposes of each process within an organism, and the starting material and end product for each.

Multiple-Choice Questions

1. A student looking through a light microscope would not be able to see any of the following cell structures EXCEPT
   (A) ribosomes
   (B) a Golgi apparatus
   (C) a nucleus
   (D) an endoplasmic reticulum
   (E) a peroxisome

2. Prokaryotic and eukaryotic cells have all of the following structures in common EXCEPT
   (A) plasma membrane
   (B) protein-complexed DNA
   (C) nucleoid region
   (D) cilia
   (E) cytoplasm

Directions: Questions 3–7 below consist of five lettered headings followed by a list of numbered phrases or sentences. For each numbered phrase or sentence, select the one heading that is most closely related to it and fill in the corresponding oval on the answer sheet. Each heading may be used once, more than once, or not at all in each group.

Questions 3–7
   (A) Peroxisomes
   (B) Golgi apparatus
   (C) Lysosomes
   (D) Endoplasmic reticulum
   (E) Mitochondria

3. An organelle that is characterized by extensive, folded membranes, and is often associated with ribosomes

4. An organelle with a *cis* and *trans* face, which acts as the packaging and secreting center of the cell

5. The sites of cell respiration

6. Single-membrane structures in the cell that perform many metabolic functions and produce hydrogen peroxide

7. Large membrane-bound structures that contain hydrolytic enzymes and that exist only in animal cells

8. Which of the following molecules is a typical component of the animal cell membrane?
   (A) Starch
   (B) Glucose
   (C) Nucleic acids
   (D) Carbohydrates
   (E) Vitamin K

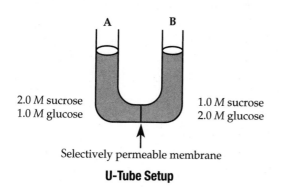

A   B

2.0 *M* sucrose
1.0 *M* glucose

1.0 *M* sucrose
2.0 *M* glucose

Selectively permeable membrane

U-Tube Setup

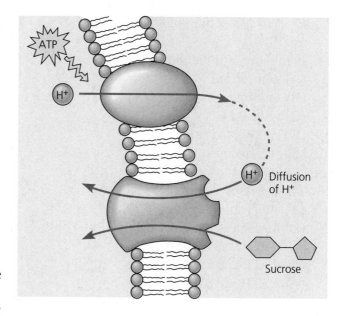

ATP

H⁺

H⁺   Diffusion
of H⁺

Sucrose

9. The drawing above shows two solutions of glucose and sucrose in a U-tube containing a semi-permeable membrane (which allows the passage of sugars). Which of the following accurately describes what will take place next?
   (A) Glucose will diffuse from side A to side B.
   (B) Sucrose will diffuse from side B to side A.
   (C) No net movement of molecules will occur.
   (D) Glucose will diffuse from side B to side A.
   (E) There will be a net movement of water from side B to side A.

10. Which of the following is an example of passive transport across the cell membrane?
    (A) The travel of an action potential down a muscle cell
    (B) The uptake of nutrients by the mircovilli of cells lining the stomach
    (C) The movement of insulin across the cell membrane
    (D) The movement of carbon dioxide across the cell membrane
    (E) The selective uptake of hormones across the cell membrane

11. The figure above illustrates the process of
    (A) cotransport
    (B) passive diffusion
    (C) receptor-mediated endocytosis
    (D) phagocytosis
    (E) pinocytosis

12. Large molecules are moved out of the cell by which of the following processes?
    (A) Pinocytosis
    (B) Phagocytosis
    (C) Receptor-mediated endocytosis
    (D) Cytokinesis
    (E) Exocytosis

13. The purpose of cell respiration in a eukaryotic cell is to
    (A) synthesize carbohydrates from CO_2
    (B) synthesize fats and proteins from CO_2
    (C) break down carbohydrates to provide energy for the cell in the form of ATP
    (D) break down carbohydrates to provide energy for the cell in the form of ADP
    (E) provide oxygen to the cell

$$2K + Br_2 \rightarrow 2K^+ + 2Br^-$$

14. In the course of the above reaction, potassium is
    (A) neutralized
    (B) oxidized
    (C) reduced
    (D) sublimated
    (E) recycled

15. The net result of glycolysis is
    (A) 4 ATP and 4 NADH
    (B) 4 ATP and 2 NADH
    (C) 2 ATP and 4 NADH
    (D) 2 ATP and 2 NADH
    (E) 4 ATP and 8 NADH

16. In the course of the Krebs cycle, how many molecules of ATP are produced?
    (A) 1
    (B) 2
    (C) 3
    (D) 4
    (E) 5

17. The process that produces the largest amount of ATP during respiration is
    (A) glycolysis
    (B) fermentation
    (C) the Krebs cycle
    (D) the electron transport chain
    (E) chemiosmosis and oxidative phosphorylation

Directions: The group of questions below consists of five lettered headings followed by a list of numbered phrases or sentences. For each numbered phrase or sentence, select the one heading that is most closely related to it and fill in the corresponding oval on the answer sheet. Each heading may be used once, more than once, or not at all in each group.

Questions 18–22
    (A) Chemiosmosis
    (B) Electron transport chain
    (C) The Krebs cycle
    (D) Glycolysis
    (E) Fermentation

18. The process by which glucose is split into pyruvate

19. The process by which a hydrogen gradient is used to create ATP

20. A process that makes a small amount of ATP and can produce lactic acid as a by product

21. A series of membrane embedded electron carriers that ultimately create the hydrogen ion gradient to drive the synthesis of ATP

22. The process by which the breakdown of glucose is completed and CO_2 is produced

23. Muscle fatigue is caused when the process of fermentation in oxygen-depleted cells produces which of the following?
    (A) ADP
    (B) Ethanol
    (C) Lactate
    (D) Uric acid
    (E) Pyruvate

24. Groups of photosynthetic pigment molecules situated in the thylakoid membrane are called:
    (A) photosystems
    (B) carotenoids
    (C) chlorophyll
    (D) grana
    (E) CAM plants

25. The main products of the light reactions of photosynthesis are
    (A) NADPH and $FADH_2$
    (B) NADPH and ATP
    (C) ATP and $FADH_2$
    (D) ATP and CO_2
    (E) ATP and H_2O

26. The process in photosynthesis that bears the most resemblance to chemiosmosis and oxidative phosphorylation in cell respiration is called
    (A) photolysis
    (B) noncyclic photophosphorylation
    (C) ATP synthase coupling
    (D) preemptive photophosphorylation
    (E) dark reaction phosphorylation

27. The reactions of the Calvin cycle are also known as the dark reactions because these reactions
    (A) occur in plants only at night
    (B) occur in dark-staining cells of plant leaves
    (C) must absorb black light from the spectrum in order to proceed
    (D) do not require light in order to proceed
    (E) take place only during the day

28. The major product of the Calvin cycle is
    (A) rubisco
    (B) oxaloacetate
    (C) ribulose bisphosphate
    (D) pyruvate
    (E) glyceraldehyde-3-phosphate

29. All of the following statements are false EXCEPT
    (A) C_3 plants grow better under hot, arid conditions than do C_4 plants

    (B) C_4 plants grow better under cold, moist conditions than do C_3 plants
    (C) C_3 plants grow better under hot, arid conditions than do CAM plants
    (D) CAM plants grow better under cold, moist conditions than do C_3 plants
    (E) CAM plants and C_4 plants both grow better under hot, arid conditions than do C_3 plants

30. All of the following statements about photosynthesis are true EXCEPT
    (A) the light reactions convert solar energy to chemical energy in the form of ATP and NADPH
    (B) the Calvin cycle uses ATP and NADPH to convert CO_2 to sugar
    (C) photosystem I contains P700 chlorophyll *a* molecules at the reaction center; photosystem II contains P680 molecules
    (D) in chemiosmosis, electron transport chains pump protons (H^+) across a membrane from a region of high H^+ concentration to a region of low H^+ concentration
    (E) the steps of the Calvin cycle are sometimes referred to as the dark reactions, because they do not require light in order to take place

Directions: The group of questions below consists of five lettered headings followed by a list of numbered phrases or sentences. For each numbered phrase or sentence, select the one heading that is most closely related to it and fill in the corresponding oval on the answer sheet. Each heading may be used once, more than once, or not at all in each group.

Questions 31–35
    (A) Telophase
    (B) Interphase
    (C) Cytokinesis
    (D) Prometaphase
    (E) Anaphase

31. Cytokinesis begins during the final stage of mitosis.

32. Division of the cytoplasm of the cell

33. Sister chromatids begin to separate.

34. The genetic material of the cell replicates to prepare for cell division.

35. Microtubules begin to attach to the centromeres of the sister chromatids.

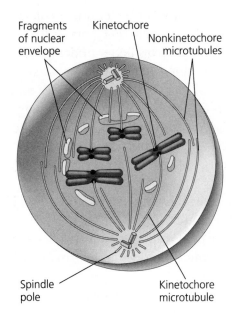

Fragments of nuclear envelope
Kinetochore
Nonkinetochore microtubules
Spindle pole
Kinetochore microtubule

36. The figure at left shows a cell in what stage of mitosis?
(A) Prophase
(B) Prometaphase
(C) Metaphase
(D) Anaphase
(E) Telophase

37. After which of the following checkpoints in the cell cycle is the cell fated to divide?
(A) G_2 phase checkpoint
(B) M phase checkpoint
(C) Interphase checkpoint
(D) G_1 phase checkpoint
(E) MPF checkpoint

Free-Response Question

1. *Prokaryotic and eukaryotic cells are physiologically different in many ways, but both represent functional collections of living matter.*

   - It has been theorized that the organelles of eukaryotic cells evolved from prokaryotes living symbiotically inside them. Compare and contrast the structure of the prokaryotic cell with eukaryotic cell organelles, and make an argument for or against this theory.
   - Trace the path of a protein in a eukaryotic cell from its creation to its excretion as waste from the cell.

ANSWERS AND EXPLANATIONS

Multiple-Choice Questions

▌ **1. (C) is correct.** Light microscopes are good for viewing objects that are about 1 mm to 100 mm in size. This means that while you can see most plant and animal cells, most bacteria, and some of the super-large organelles in the cell (like the nucleus and perhaps some mitochondria) with a light microscope, you'll need an electron microscope to see the other organelles listed. In the list of cell structures, the largest one was the nucleus. Light microscopes have lower magnification than electron microscopes.

▌ **2. (C) is correct.** The only answer choice that lists a cell structure not found in both prokaryotes and eukaryotes is *C*, nucleoid region. While eukaryotic cells have a true nucleus, which is surrounded by a membrane called a nuclear envelope, the genetic material of prokaryotes is localized in a clump in one particular region of the cell (but it is not enclosed by a membrane). Both prokaryotes and eukaryotes contain a plasma membrane, cytoplasm, DNA that's associated with proteins, and cilia (although not *all* eukaryotes and prokaryotes possess all of the cell structures listed).

3. **(D) is correct.** The endoplasmic reticulum (ER) is an organelle characterized by extensive, folded membranes, and it is often associated with ribosomes. It is the site of protein synthesis, since the ribosomes associated with rough ER are protein-producing.

4. **(B) is correct.** The Golgi apparatus is the organelle that has a *cis* and *trans* face, and it acts as the packaging and secreting center of the cell. It consists of a series of flattened sacs of membranes called cisternae.

5. **(E) is correct.** Mitochondria are the powerhouses of the cell; cell respiration takes place in the mitochondria to create ATP, the cell's energy currency. They are bound by double membranes, and the proteins involved in ATP production are embedded in the inner membranes of the mitochondria.

6. **(A) is correct.** Peroxisomes perform many metabolic functions in the cell, including the production of hydrogen peroxide.

7. **(C) is correct.** Lysosomes are characteristic of animal cells but not plant cells. They are large membrane-bound structures that contain hydrolytic enzymes, and they are responsible for the breakdown of proteins, polysaccharides, fats, and nucleic acids. They function best at a low pH (around 5), so they pump hydrogen ions from the cytosol into their lumen to maintain this acidic pH.

8. **(D) is correct.** The only answer choice listed that names a molecule typically found in the plasma membranes of animal cells is *D*, carbohydrates. The major components of animal cell membranes are phospholipids, integral and peripheral proteins, and carbohydrates. One of the main functions of carbohydrates at the cell membrane is cell-cell recognition, so carbohydrates are an important component of the immune system. Cell surface carbohydrates are unique to each organism.

9. **(D) is correct.** Substances will move down their concentration gradient until their concentration is equal on two sides of a membrane. For this reason, because the concentration of glucose on side B of the tube is 2.0 *M*, while the concentration of glucose on the A side of the tube is 1.0*M*, glucose will move to side A.

10. **(D) is correct.** The only substance listed that can passively diffuse through the cell membrane is carbon dioxide. Remember that passive diffusion is diffusion that occurs without the cell doing any work, and that the substances that can passively diffuse across the membrane are small nonpolar molecules, such as carbohydrates, carbon dioxide, and oxygen. Other substances need the processes of facilitated diffusion and the help of transport proteins to cross the membrane. This is true of all of the answer choices listed besides *D*.

11. **(A) is correct.** This figure illustrates the process of cotransport. In cotransport, a pump that is powered by ATP transports a specific solute—in this case protons—out of the cell. The protons then travel down their concentration

gradient back into the cell, passing through another transport protein and indirectly providing energy for the movement of another substance (glucose in this case) against its concentration gradient and out of the cell.

12. **(E) is correct.** Large molecules are moved out of the cell by exocytosis. In exocytosis, vesicles that are to be exported from the cell (usually coming from the Golgi apparatus) fuse with the plasma membrane, and their contents are expelled into the extracellular matrix. Pinocytosis, phagocytosis, and receptor-mediated endocytosis are types of endocytosis; cytokinesis is cell division.

13. **(C) is correct.** The purpose of cell respiration in eukaryotes is to create energy to perform work in the cell in the form of ATP. Respiration is an aerobic process, meaning that it requires oxygen in order to take place. Answer choices A and B are incorrect, because respiration involves the breakdown (not the synthesis) of carbohydrates and fats and proteins. Choice D is wrong because ADP is the product of the dephosphorylation of ATP—it is left over after the energy from ATP has been used.

14. **(B) is correct.** In the course of the reaction shown, potassium (K) is oxidized. Oxidation involves the loss of an electron. Reduction is the gain of an electron by an atom or molecule. In this reaction, potassium is the reducing agent; it reduces bromide and becomes oxidized. Bromide is the oxidizing agent; it becomes reduced when it receives an electron from potassium.

15. **(D) is correct.** The net energy result of glycolysis is the production of two molecules of ATP and two molecules of NADH. Glycolysis is the first of the three stages of respiration—the second being the Krebs cycle and the third being chemiosmosis/oxidative phosphorylation. During glycolysis, glucose is broken down and oxidized to form pyruvate. Glycolysis occurs in the cytosol, and the pyruvate it produces travels to the mitochondria to take part in the Krebs cycle.

16. **(D) is correct.** In the Krebs cycle, 4 ATP are produced. The Krebs cycle takes in a molecule called acetyl-CoA (pyruvate is converted into acetyl-CoA before it enters the Krebs cycle), and this is joined to a four-carbon molecule of oxaloacetate to form a six-carbon compound citrate that is then broken down again to produce oxaloacetate; the oxaloacetate reenters the cycle. In the course of the Krebs cycle, the following are produced: 4 CO_2, 2 ATP, 6 NADH and 2 FADH.

17. **(E) is correct.** The process that produces the most ATP during cell respiration is chemiosmosis and oxidative phosphorylation. The total ATP production by each of the phases is:

Glycolysis: 2ATP

Fermentation: 2 ATP (both lactic acid fermentation and alcohol fermentation produce 2 ATP)

The Krebs cycle: 1 ATP

The electron transport chain: Produces no net ATP by itself, but creates a H^+ gradient that drives the production of ATP

Chemiosmosis and oxidative phosphorylation: FADH$_2$ and NADH donate electrons to the electron transport chain, which is coupled to ATP synthesis by the process called chemiosmosis; the movement of electrons down the electron transport chain creates a H$^+$ gradient across the mitochondrial membrane, which drives the synthesis of ATP from ADP.

18. (D) is correct. In glycolysis, glucose is oxidized to two molecules of pyruvate. This is the first step in cell respiration, and it also produces 2 ATP molecules and 2 NADHs.

19. (A) is correct. In chemiosmosis, the hydrogen gradient created by the transfer of electrons in the electron transport system provides the power to synthesize ATP from ADP.

20. (E) is correct. Fermentation is an anaerobic alternative to cell respiration. It consists of glycolysis and a bunch of reactions that serve to regenerate NAD$^+$; electrons are transferred from NADH to pyruvate or its derivatives, and the NAD$^+$ oxidizes sugar in glycolysis. There are two main types of fermentation—alcohol fermentation (which creates ethanol as a product) and lactic acid fermentation (which creates lactate).

21. (B) is correct. The electron transport chain is a series of inner mitochondrial matrix membrane-embedded molecules that are capable of being oxidized and reduced as they pass along electrons. The energy produced from the passage of these electrons down the chain is used to create an H gradient across the membrane, and the flow of H$^+$ ions down the gradient and back across the membrane powers the phosphorylation reaction of ADP to form ATP.

22. (C) is correct. The Krebs cycle contains the final reactions for the breakdown of glucose that was started in glycolysis. The pyruvate from glycolysis is converted into acetyl CoA, which enters the cycle and is joined to oxaloacetate to create citrate, which is then converted to oxaloacetate again and reused. This cycle gives off CO$_2$, and forms 1 ATP, 3 NADH, and 1 FADH. The cycle goes through one rotation to break down each of the molecules of pyruvate produced in glycolysis (which of course is converted to acetyl CoA), so the net results of the breakdown of one glucose molecule is 2 ATP, 6 NADH, and 1 FADH.

23. (C) is correct. When muscle cells in the body are depleted of oxygen, they switch from cell respiration to lactic acid fermentation. In lactic acid fermentation, NADH reduces pyruvate directly, and lactate is formed as a waste product. This lactic acid fermentation route is taken when exercise is just begun, and all of the cell's metabolic machinery is using the oxygen to break down sugars in the cell. The lactate that builds up in the cell as the waste product from fermentation is broken down again when it is filtered through the liver.

24. (A) is correct. Groups of photosynthetic pigment molecules in the thylakoid membrane are called photosystems. There are two photosystems involved in photosynthesis, photosystem I and photosystem II. Both contain chlorophyll molecules and many proteins and other organic molecules, and both have an antenna-complex that is able to harness incoming light. Each of these photosystems contains a reaction center, where chlorophyll a and the primary electron acceptor are located. Photosystem II is the site of the first redox reaction of photosynthesis.

■ 25. **(B) is correct.** The main products of the light reactions of photosynthesis are NADPH and ATP. This ATP and NADPH is used to convert CO_2 to sugar in the Calvin cycle; the enzyme rubisco combines the CO_2 with ribulose bisphosphate (RuBP), and electrons from NADPH and energy from ATP are used to synthesize a three-carbon molecule called glyceraldehyde-3-phosphate—some of which leaves the cycle and is used to make carbohydrates for the cell.

■ 26. **(B) is correct.** The process in the dark reactions of photosynthesis that bears the closest resemblance to chemiosmosis and oxidative phosphorylation in cell respiration is noncyclic photophosphorylation. In this process, energy from the transfer of electrons down the electron transport chain embedded in the thylakoid membrane wall. Later this ATP is used as energy in the formation of carbohydrates in the Calvin cycle.

■ 27. **(D) is correct.** The Calvin cycle reactions are also called the dark reactions, because they do not need light in order to proceed. In fact, the Calvin cycle works mainly during the day in plants, because the products of the light reactions (which are necessary for the Calvin cycle to proceed) are created during the day, and so this is when they are available to enter the dark reactions. The two sets of reactions are coordinated in the chloroplast.

■ 28. **(E) is correct.** The organic product of the Calvin cycle, which is later used to build large carbohydrates in the cell, is glyceraldehyde-3-phosphate, or G3P. This molecule is created as a result of the fixation of three molecules of CO_2, which costs the cell ATP and NADPH that was created in the light reactions of photosynthesis.

■ 29. **(A) is correct.** C_4 and CAM plants both grow better than do C_3 plants under conditions of increased median air temperature and decreased relative humidity. Both C_4 and CAM plants utilize an alternative method for photosynthesis that enables them to fix carbon more quickly even when there is little of it around.

■ 30. **(D) is correct.** The electron transport chains pump protons across membranes from regions of *low* H^+ concentrations to regions of *high* H^+ concentrations; all of the other answer options are true. This proton pumping occurs in both mitochondria and chloroplasts, and the protons then diffuse (with the concentration gradient) back across the membrane (through ATP synthetase); this drives the synthesis of ATP.

■ 31. **(A) is correct.** In telophase, nuclear envelopes begin to form again around the sets of chromosomes, which are now located at opposite ends of the cell. The chromatin becomes less condensed, and cytokinesis begins—the cytoplasm of the cell is divided.

■ 32. **(C) is correct.** During cytokinesis, the cytoplasm of the cell is divided approximately equally as the cell membrane pinches off (in animal cells) to create two daughter cells; a cell plate forms in plant cells.

■ 33. **(E) is correct.** In anaphase, the sister chromatids, which were lined up along the equator of the cell, begin to separate, pulled apart by the retracting microtubules. By the end of anaphase, the opposite ends of the cell contain complete and equal sets of chromosomes.

■ **34. (B) is correct.** Interphase is not a part of mitosis; rather it is the part of the cell cycle when the cell gets ready to divide by replicating its DNA. There are three stages in interphase—G_1 phase, S phase, and G_2 phase.

■ **35. (D) is correct.** Prometaphase is the phase of mitosis in which the nuclear envelope begins to fragment so that the microtubules can begin to attach to the kinetochores of the chromatids, which by this time are very condensed.

■ **36. (B) is correct.** The depicted cell is in prometaphase. As you can see, the nuclear envelope is fragmenting, and the microtubules have already attached to some of the kinetochores at the centromeres of the chromosomes. The chromosomes are condensed and beginning to move so that they are lined up along the cell's equator.

■ **37. (D) is correct.** The most crucial checkpoint of the cell cycle is the G_1 checkpoint. In the cell cycle, a checkpoint is a point at which there can be a signal to stop or to go ahead. If a cell receives the signal to go ahead at the G_1 checkpoint, it will complete the cycle and divide. If it does not receive the go-ahead signal, it will enter the (nondividing) G_o phase for an indeterminate period of time.

Free-Response Question

1. Eukaryotic cell organelles might have evolved from free-living prokaryotic organisms. First of all, prokaryotic cells are much smaller than eukaryotic cells—they range from 100 nm to 100 μm, compared to the average size of eukaryotes: 10 μm – 100 μm. However, mitochondria (organelles unique to eukaryotic cells, and functioning in the creation of ATP in cell respiration) and eukaryotic cell nuclei are comparable in size to prokaryotic cells, ranging from about 1 mm – 10 mm. This is one piece of possible evidence for their having an evolutionary origin like certain types of prokaryotic cells.

Another interesting characteristic of organelles that may tie them to prokaryotes is their structure and cell contents. To illustrate this, let's look at the structure of the mitochondria. Mitochondria are found in all animal cells, plant cells, fungi, and protists. They can exist in great numbers in these cells, or cells can contain just one mitochondria (depending on the metabolic activity of the cell). It has been observed that mitochondria can move around, alter their shape, and even divide in two—all of which are characteristic of living cells. Their structure consists of a double membrane exterior (the membrane is a typical combination of phospholipids and proteins, like the membrane of the cell itself); the outer membrane is relatively smooth, but the interior membrane has infoldings called cristae. This creates two different compartments in mitochondria: the inner compartment is the mitochondrial matrix, and the compartment in-between the two membranes is called the intermembrane space. Mitochondria also contain mitochondrial DNA. Not very much DNA is contained in mitochondria, but the presence of DNA seems like evidence of their having been free-standing organisms at one time. Also similarly to mitochondria, prokaryotes do not contain many interior structures other than their genetic material (which is not enclosed in a nucleus) and their cell membranes. All of the above indicates a close evolutionary relationship between mitochondria and prokaryotic cells.

In order to trace the path of proteins in the cell, from their creation to their expulsion, we must start in the nucleus. The nucleus can be thought of as the brain of the cell, because it contains the cell's genetic material in the form of DNA. In the nucleus, mRNA is created from the transcription of DNA, and mRNA travels out of the nucleus to the cytoplasm, ending up at ribosomes, which are associated with the endoplasmic reticulum (called rough endoplasmic reticulum because of this association). Here they are translated into proteins, which then undergo folding to assume their final shape, or conformation. Proteins then either carry out their metabolic function in the cell, whether they act as structural components, enzymes, etc, or they are packaged for secretion from the cell. We will discuss the latter fate of a (secretory) protein to show how it is exported from the cell.

Secretory proteins travel from the endoplasmic reticulum to the series of flattened membranous sacs knows as the Golgi apparatus. They enter at the *cis* face, where they bud off in vesicles from fold to fold through the length of the Golgi, to eventually bud off the *trans* face, after undergoing a series of modifications to prepare them for secretion. The vesicles may then fuse with the cell membrane, and the contents are endocytosized to the cell's exterior.

This response shows that the writer used the following key terms in context, showing the writer's knowledge of their meanings and relatedness:

organelles	*mRNA*
prokaryote	*ribosomes*
eukaryote	*endoplasmic reticulum*
mitochondria	*conformation*
metabolism	*enzymes*
phospholipids	*secretory proteins*
proteins	*Golgi apparatus*
cristae	*cis/trans face*
mitochondrial matrix	*vesicle*
DNA	

The response also contains an explanation of the following subjects and processes:

    —size comparison of prokaryotes and eukaryotes (with figures)
    —structure, content, and behavior of mitochondria vs. prokaryotes
    —location of transcription and translation
    —pathway of proteins

The response also is well organized. Each of the separate points are raised in a new paragraph, and the paragraphs consistently have thesis sentences and conclusions.

CHAPTER 3

Genetics

Meiosis and Sexual Life Cycles: An Introduction to Heredity

▌ **Heredity** is defined as the transmission of traits from one generation to the next. This is also known as **inheritance**.

▌ **Genetics** is the study of heredity and variation. **Variation** is the genetic difference between siblings or members of the same species.

▌ **Genes** are DNA segments—the basic units of heredity that are transmitted from one generation to the next.

▌ The location of a gene on the chromosome is called its **locus** (and the plural of *locus* is *loci*).

▌ **Asexual reproduction** is a form of reproduction in which a single parent is involved in passing on all of its genes to its offspring. Some organisms capable of reproducing asexually are some single-celled eukaryotes and the hydra (which is related to the jellyfish). Asexually-produced organisms are similar in appearance.

▌ **Sexual reproduction** is a form of reproduction in which two individuals (parents) are contributing genes. This form of reproduction results in greater genetic variation in the offspring.

The Role of Meiosis in Sexual Life Cycles

▌ The **life cycle** of an organism is defined as the sequence of stages in its reproductive history through the course of one generation.

▌ **Somatic cells** are cells other than gametes, or egg and sperm cells. They are the body cells of an organism. Each somatic cell in humans has 46 chromosomes.

▌ The **karyotype** of an organism refers to a picture of its complete set of chromosomes, arranged in pairs of homologous chromosomes from the largest size pair to its smallest size pair. **Homologous chromosomes** are those that carry the genes that control the same traits. They are similar in length and in the position of their centromere, and they have the same staining pattern. One homologous chromosome from each pair is inherited from each parent; in other words, half of the set of 46 chromosomes in your somatic cells was inherited from your mother (she donated 23) and the other half (23) was donated by your father.

▌ Exceptions to the rule that all chromosomes are part of a homologous pair are the **sex chromosomes**—in humans, it is the X and Y. Human females have a homologous pair of chromosomes, XX, but males have one X chromosome and one Y chromosome. The non-sex chromosomes are called **autosomes**.

- **Gametes**—meaning sperm and ovum (egg)—are **haploid** cells. That is, they contain half the number of chromosomes of somatic cells. They contain 22 autosomes plus a single sex chromosome (either X or Y, giving them a haploid number), giving them a haploid number of 23. The haploid number of chromosomes can be symbolized by n.
- In **fertilization** (a combination of a sperm cell and an egg cell), haploid gametes from the parents fuse, and the fertilized egg that results is called a **zygote**. It is diploid (has two sets of chromosomes), as are all the somatic cells of an organism, which are derived from the zygote. The diploid number of chromosomes is symbolized by $2n$.
- **Meiosis** is the process by which, in the course of gamete production, the chromosome number is halved so that haploid gametes are formed. Fertilization restores the diploid number as the gametes are combined. Fertilization and meiosis alternate in the life cycles of sexually reproducing organisms. There are three types of life cycles:

  1. *In humans and most animals*, meiosis occurs during gamete production, and the diploid zygote divides by mitosis to produce a diploid multicellular organism.
  2. *In fungi, some protists, and algae*, after gametes fuse to form the diploid zygote, meiosis occurs to produce haploid cells. These then divide by mitosis to give a haploid multicellular organism.
  3. *In plants and some algae*, **alternation of generations** occurs, including a haploid and diploid stage in the life cycle. The diploid stage is the sporophyte, and meiosis in the diploid phase creates haploid spores, which divide mitotically to produce a gametophyte. The gametophyte produces haploid gametes through mitosis, and the fertilization occurs, producing a diploid zygote.

Meiosis Reduces Chromosome Number from Diploid to Haploid

- Meiosis and mitosis look similar—both are preceded by the replication of the cell's DNA, for instance, but in meiosis this replication is followed by *two* stages of cell division, **meiosis I** and **meiosis II**.
- The result of meiosis is four daughter cells, each of which has half as many chromosomes as the parent cell.

The stages of meiosis are as follows:

- *Interphase*: Each of the chromosomes replicate, resulting in two sister chromatids attached at their centromeres. The centrosomes also replicate in this phase.
- *Prophase I*: The chromosomes condense, and the homologues (consisting of two sister chromatids) pair up. **Synapsis** occurs—that is, the joining of two pairs of homologous chromosomes along their length. This newly formed structure is called a **tetrad**. It has four chromotids. Parts of the homologous chromosomes undergo **crossing over** at **chiasmata** (places at which homolo-

gous chromosomes overlap during synapsis). The centrioles move away from each other; the nuclear envelope disintegrates; and spindle microtubules attach to the kinetochores forming on the chromosomes that begin to move to the metaphase plate of the cell.

▌ *Metaphase I*: The homologous pairs of chromosomes are lined up at the metaphase plate, and microtubules from each pole attach to each of the members of the homologous pairs, in preparation for pulling them to opposite ends of the cell.

▌ *Anaphase I*: The spindle apparatus helps to move the chromosomes toward opposite ends of the cell; sister chromatids stay connected and move together toward the poles.

▌ *Telophase I, Cytokinesis*: The homologous chromosomes move until they reach the opposite poles, so that each pole contains a haploid set of chromosomes, with each chromosome still made up of two sister chromatids. Cytokinesis occurs at the same time as telophase—a cleavage furrow occurs in animal cells, and cell plates occur in plant cells. Both result in the formation of two daughter cells.

▌ *Prophase II*: A spindle apparatus forms, and sister chromatids move toward the metaphase plate.

▌ *Metaphase II*: The chromosomes are lined up on the metaphase plate, and the kinetochores of each sister chromatid prepare to move to opposite poles of the cell.

▌ *Anaphase II*: The centromeres of the sister chromatids separate, and individual chromosomes move to opposite ends of the cell.

▌ *Telophase II and Cytokinesis*: The chromatids have moved all the way to opposite ends of the cell; nuclei reappear, and cytokinesis occurs. Each daughter cell (there are 4 total) has the haploid number of chromosomes.

Origins of Genetic Variation

Here are some processes that contribute to variation in the offspring of sexually reproducing organisms:

▌ *Independent Assortment of Chromosomes*—In metaphase I, when the homologous chromosomes are lined up on the metaphase plate, they can pair up in any combination, with any of the homologous pairs facing either pole. This means that there is a 50–50 chance that a particular daughter cell will get a maternal chromosome or a paternal chromosome from the homologous pair.

▌ *Crossing Over*—After prophase I, homologous chromosomes synapse and the homologous chromosomes exchange homologous parts of two non-sister chromatids. Then, during metaphase II, chromosomes that now have recombinant chromatids can be facing either of two poles with respect to each other, which further increases variation in reproduction.

▌ *Random Fertilization*—This refers to the fact that fertilization (in which an egg meets with a sperm) is random. Since each egg and sperm is different, as a result of independent assortment and crossing over, each combination of egg and sperm is unique.

Mendel and the Gene Idea

In this section, the most important thing for you to understand is not the history of Gregor Mendel, his work and theories, but rather the *results* of his work and theories. The AP Biology Exam often contains questions asking what offspring will result from certain combinations of parents—and their genes.

Gregor Mendel's Discoveries

▊ In genetics, the word *trait* describes a heritable feature of an organism (such as eye color) that varies among individuals.

▊ A **variation** is the difference in a trait—for example, in humans blue eyes or brown eyes is a trait.

▊ **Pure** or **true-breeding** is a term used by Mendel in his experiments. It refers to a phenomenon such as this one: when a pea plant is self-pollinated, all of the offspring are of the same type. In other words, a white pea plant would give rise only to white pea plant offspring. The crossing (or mating) of two true-breeding varieties of an organism is called **hybridization**.

▊ The true-breeding parents in a hybridization are called the **P (parental) generation**; their offspring are called the F_1 **(first filial) generation**. If the F_1 population are crossed, their offspring are called the F_2 **(second filial) generation**.

Below is a brief summary of the four major conclusions that Mendel made in the course of his experiments.

▪ *Alternative versions of genes cause variations in inherited characteristics among offspring.* For example, consider flower color in peas. The gene for flower color in pea plants comes in two versions—white and purple. These alternative versions of the gene, called **alleles**, are the result of slightly different DNA content at homologous loci on chromosomes.

▪ *For each character, every organism inherits one allele from each parent.*

▪ *If the two alleles are different, then the **dominant allele** will be fully expressed in the offspring, while the **recessive allele** will have no noticeable effect on the offspring.*

▪ *The two alleles for each character separate during gamete production.* If the parent has two of the same alleles, then the offspring will all get that version of the gene, but if the parent has two different alleles for a gene, each offspring has a 50/50 chance of getting one of the two alleles. This is Mendel's **law of segregation**.

▊ Another of Mendel's laws is the **law of independent assortment.** It states that each pair of alleles will segregate (separate) independently during gamete formation.

Below are some terms to memorize for the exam.

1. The term **homozygous** refers to an organism that has two identical alleles for a particular trait. If the dominant allele for a trait is designated as *R* (dominant

traits are generally capitalized), and the recessive allele is designated *r* (recessive traits are generally not capitalized), then an individual homozygous for the dominant trait would be *RR*

2. A **heterozygous** organism has two different alleles for a trait.
3. **Phenotype** refers to an organism's expressed physical traits.
4. **Genotype** refers to an organism's genetic makeup.
5. **Testcross** refers to the crossing of a recessive homozygote with an individual exhibiting the dominant phenotype, in order to find out if the organism is homozygous-dominant or heterozygous-dominant.
6. A **monohybrid cross** is a cross involving the study of only one character, i.e. flower color.
7. A **dihybrid cross** is a cross intended to study two characters, i.e. seed color and seed shape.

The following diagram shows the results of a monohybrid cross between two plants: one with white flowers that is homozygous recessive (*pp*), and one with purple flowers that is homozygous dominant (*PP*). Across the top are the possible gametes produced by one plant, and along the side are the gametes produced by the other plant parent.

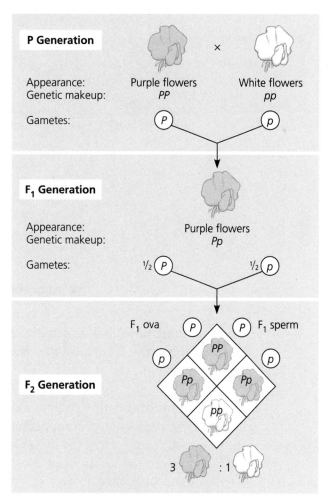

Mendel's Law of Segregation

The diagram below shows the results of a dihybrid cross. In this case, in the parental generation we are crossing two homozygous plants: one homozygous dominant for purple flowers and yellow seeds (*PPYY*) and one homozygous recessive for white flowers and green seeds (*ppyy*). The only gamete type the first can produce is *PY*, and the only gamete the second can produce is *py*. The F$_1$ generation, therefore, is composed of individuals with genotype *PpYy*. This gene combination or genotype *PpYy* is called a **dihybrid genotype**. Crossing *PpYy* gives an F$_2$ that looks like this:

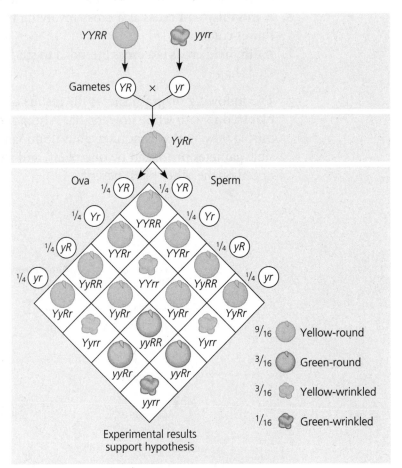

Hypothesis: Independent Assortment

Understanding how to work genetic crosses involves familiarity with the basic laws of probability. There are two laws that you will use directly in working genetics problems.

▌ **The rule of multiplication:** When calculating the probability that two or more independent events will occur together in a specific combination, you multiply the probabilities of each of the two events. Thus, the probability of a coin landing face up two times in two flips is:

$$\frac{1}{2} \times \frac{1}{2} = \frac{1}{4}.$$

- **The rule of addition**: When calculating the probability of an event that could happen in any of a number of ways, you add the probabilities of the ways it could happen. If tossing a coin, for example, which side would come up first—*tail and head* or *head and tail*?

$$(\frac{1}{2} \times \frac{1}{2}) + (\frac{1}{2} \times \frac{1}{2}) = \frac{1}{4} = \frac{1}{2}.$$

Extending Mendelian Genetics

- **Incomplete dominance** is a type of dominance in which the F_1 hybrids have an appearance that is in-between that of the two parents. For instance, if a plant with white flowers were crossed with a plant with red flowers, and if all of the offspring had pink flowers, that is an example of incomplete dominance. Breeding two of the hybrids with incomplete dominance gives a ratio of 1:2:1, red:pink:white.
- **Complete dominance** is dominance in which the heterozygote and the homozygote for the dominant allele are indistinguishable.
- **Codominance** occurs when two alleles are dominant and affect the phenotype in two different and equal ways. The traditional example for this type of dominance is the human blood type.
- Most genes exist in different forms—that is, **multiple alleles**. Again, a good example of this is seen in human blood typing. The chart below will help you familiarize yourself with blood types.

Phenotype	Genotype	Antibodies Expressed
A	$I^A I^A$ or $I^A i$	Anti-B
B	$I^B I^B$ or $I^B i$	Anti-A
AB	$I^A I^B$	None
O	ii	Anti-A, Anti-B

- **Pleiotropy** is the ability of a gene to affect many different traits in an organism.
- **Epistasis** refers to the ability of a gene at one location to alter the effects of a gene at a distant location.
- In **polygenic inheritance**, two or more genes have an additive effect on a single character in the phenotype.

Mendelian Inheritance in Humans

- A pedigree is a diagram—basically a family tree—that describes the relationship between parents and offspring across two or more generations. In a typical pedigree, circles represent women, and squares represent men. White open circles or squares indicate that the individual did not express a particular trait, whereas black indicates that the individual expressed that trait. Through the patterns they reveal, pedigrees can help determine the genome of individuals that make them up; pedigrees can also help predict the genome of future offspring.

- Genes causing **recessively inherited disorders** are inherited like any other recessive genes. Alleles that cause genetic disorders either code for a dysfunctional protein or for no protein at all. Heterozygotes that inherited one copy of the mutant allele and one copy of the normal allele usually have a normal phenotype, because the normal allele can usually produce enough protein to make up for the mutant allele's function.
- Heterozygotes with the normal phenotype but carrying one mutant allele are called **carriers**.
- **Cystic fibrosis** is caused by a mutation in an allele that codes for a certain cell membrane protein that functions in the transport of chlorine into and out of the cell.
- **Tay-Sachs disease** is caused by an allele that codes for a dysfunctional enzyme that is unable to break down certain lipids in the brain.
- **Sickle cell disease** is caused by an allele that codes for a mutant hemoglobin molecule that forms long rods when the oxygen levels in the blood are low.
- Lethal dominant alleles are a lot less common than lethal recessive alleles. This is because the lethal allele usually kills offspring before they are mature and can reproduce, so the lethal allele won't be passed on to the next generation.
- Only late-acting lethal alleles are passed on. One example of this is the **Huntington's disease** allele.

The Chromosomal Basis of Inheritance: Relating Mendelism to Chromosomes

- In the early 1900s, the **chromosome theory of heredity** was formed. It stated that genes have specific locations (called loci) on chromosomes, and that it is chromosomes that segregate and assort independently.
- After the chromosome theory of heredity was formed, Thomas Hunt Morgan discovered a sex-linked gene. A **sex-linked gene** is one located on a sex chromosome (X or Y in humans). Non sex-linked genes found on non-sex chromosomes are called **autosomes**.
- **Linked genes** are those located on the same chromosome, and thus they are inherited together during cell division. An example in humans is the linkage of the red-haired gene with the freckles gene on the same chromosome.
- **Genetic recombination** is the production of offspring with a new combination of genes inherited from the parents.
- **Recombinants** are individuals who receive new combinations of genes from their parents. **Parental types** receive nonrecombinant genes, and their phenotype matches that of one of the parents.
- During meiosis, unlinked genes follow independent assortment because they are located on different chromosomes. Linked genes are located on the same chromosome and would not seem to follow independent assortment. However, recombinations of linked genes are explained by crossing over. The further apart two genes are on a chromosome, the greater their chance of crossing over—thus, the greater their chance of assorting independently.
- Geneticists use recombination data to construct a **genetic map**, which is an ordered list of the genes and their loci along a particular chromosome.

- A **linkage map** is a genetic map that is based on recombination frequencies, and map units are used to express distances along the chromosome. One map unit is equal to a 1% recombination frequency.

Sex Chromosomes

- In humans, there are two types of sex chromosomes, X and Y. Normal females have two X chromosomes, whereas normal males have one X and one Y chromosome.
- In the testes and ovaries, the sex chromosomes segregate like any other chromosome pair into separate gametes during meiosis. Each ovum contains an X chromosome; there are two types of sperm—those with an X chromosome and those with a Y chromosome. In fertilization, there is a 50/50 chance that a sperm carrying an X or Y will reach and penetrate the egg first. Thus, gender is determined by chance and by the male sperm cell in humans.
- Sex-linked genes carry genes for many characters that are not related to sex. This is especially true of the X chromosome, so the term "sex-linked" is usually used for genes that are found on the X chromosome, not the Y.
- Fathers pass sex-linked genes on to their daughters but not to their sons.
- Females will express a sex-linked trait only if they are homozygous for it, but because males have only one X chromosome, if they have the sex-linked gene, there will be no other normal X chromosome allele to mask the effects of the mutant, and so they will express the trait.

Some sex-linked disorders include:

- **Duchenne muscular dystrophy**—characterized by a progressive weakening of the muscle, caused by the absence of a muscle protein called dystrophin
- **Hemophilia**—characterized by having blood with an inability to clot normally, caused by the absence of proteins required for blood clotting
- While female mammals inherit two copies of the X chromosome, one of the X chromosomes (randomly chosen) in each cell of the body becomes inactivated during embryonic development. This evens out the dose of genes that females get, compared to males.
- The inactivated chromosome condenses into a **Barr body**, which associates with the nuclear envelope. Still, females are not affected as heterozygote carriers of problematic alleles, because half of their autosomes are normal and are producing the necessary protein.

Errors and Exceptions in Chromosomal Inheritance

- When the members of a pair of homologous chromosomes do not separate properly during meiosis I, or sister chromatids don't separate properly during meiosis II, **nondisjunction** occurs.
- As a result of nondisjunction, one gamete receives two copies of the gene, while the other gamete receives none. In the next step, if the faulty gametes engage in fertilization, the offspring will have an incorrect chromosome number. This is known as **aneuploidy**.

- Fertilized eggs that have received three copies of the chromosome in question are said to be **trisomic;** those that have received just one copy of a chromosome are said to be **monosomic** for the chromosome.
- If nondisjunction occurs during mitosis (and early in embryonic development), it will be passed on to a large number of the organism's cells and have a significant effect on the organism.
- **Polyploidy** is the condition of having more than two complete sets of chromosomes, and this is somewhat common in plants.

Below are some common chromosome structure alterations that you should review.

- **Deletion** refers to a chromosome segment that has no centromere. It is broken off and lost during segregation. The cell that receives the partial chromosome will be missing all of the genes located on the chromosome fragment.
- If the chromosome fragment that broke off (causing the deletion above) becomes attached to its sister chromatid, this causes a **duplication**. In this case, the zygote will get a double-dose of the genes located on that chromosome.
- An **inversion** refers to a chromosome fragment breaking off and then reattaching to its original position—but backwards, so that the part of the fragment that was originally at the attachment point is now at the end of the chromosome.
- A **translocation** occurs when the chromosome fragment joins a nonhomologous chromosome. This moves a segment of one chromosome to a nonhomologous chromosome. A translocation can be reciprocal—that is, the nonhomologous chromosome can exchange segments.

Human disorders caused by chromosome alterations include:

- **Down syndrome**—an aneuploid condition that is the result of having an extra chromosome 21 (People affected are trisomic for 21)
- Klinefelter syndrome—an aneuploid condition in which a male possesses the sex chromosomes XXY (an extra X)
  - Turner syndrome—a condition in which those affected have just one sex chromosome, an X (This is called **monosomy**)

The Molecular Basis of Inheritance: DNA as the Genetic Material
- **X-ray crystallography** is a process used to visualize molecules three-dimensionally. X-rays are diffracted as they pass through the molecule, and they bounce back to produce patterns that can be interpreted through mathematical equations. Through this technique the structure of DNA was first visualized.
- DNA is a **double helix**, which can be described as a twisted ladder with rigid rungs. The side or backbone is made up of a sugar-phosphate components, whereas the rungs are made up of pairs of nitrogenous bases:

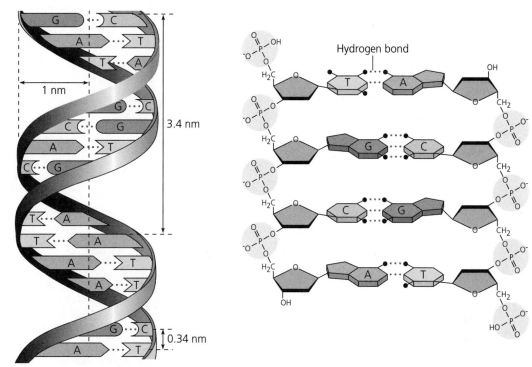

Key Features of DNA Structure

Partial Chemical Structure

▌ The nitrogenous bases of DNA are **adenine** (**A**), **thymine** (**T**), **guanine** (**G**), and **cytosine** (**C**). In DNA, adenine pairs only with thymine, and guanine pairs only with cytosine. Adenine and thymine are able to form two hydrogen bonds between them in DNA, and cytosine and guanine form three hydrogen bonds.

DNA Replication and Repair

▌ DNA replication is **semi-conservative**. This means that at the end of its replication, each of the daughter molecules has one old strand, derived from the parent strand of DNA, and one strand that is newly synthesized.

▌ The replication of DNA begins at sites called the **origins of replication**.

▌ Initiation proteins bind to the origin of replication and separate the two strands to create a replication bubble. DNA replication then proceeds in both directions along the DNA strand until the molecule is copied.

▌ An enzyme called **DNA polymerase** catalyzes the elongation of new DNA at the replication fork, but there are a dozen or more enzymes and other proteins also involved in the process of DNA replication.

▌ DNA polymerase adds nucleotides to the growing polypeptide chain one by one, and each nucleotide loses two phosphate groups, which are subsequently hydrolyzed in an exergonic reaction to fuel the process of polymerization.

▌ The strands of DNA are antiparallel, meaning that their sugar phosphate backbones run in opposite directions. This means that DNA replication occurs continuously along one strand, which is called the **leading strand**, and discontinuously along the other strand, the **lagging strand**.

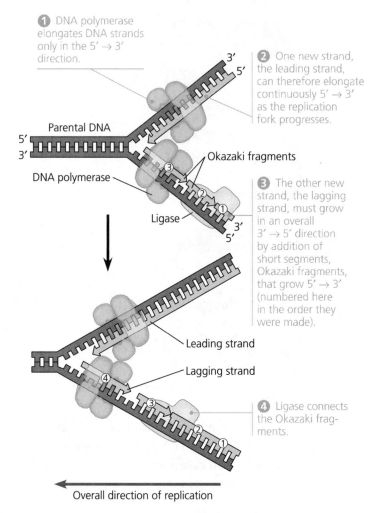

1 DNA polymerase elongates DNA strands only in the 5′ → 3′ direction.

Parental DNA

DNA polymerase

Ligase

Okazaki fragments

2 One new strand, the leading strand, can therefore elongate continuously 5′ → 3′ as the replication fork progresses.

3 The other new strand, the lagging strand, must grow in an overall 3′ → 5′ direction by addition of short segments, Okazaki fragments, that grow 5′ → 3′ (numbered here in the order they were made).

Leading strand

Lagging strand

4 Ligase connects the Okazaki fragments.

Overall direction of replication

Synthesis of Leading and Lagging Strands during DNA Replication

▌ The lagging strand is synthesized in separate pieces called **Okazaki fragments**, which are then sealed together by **DNA ligase** to create a regular DNA strand.

▌ An enzyme called **primase** is responsible for initiating DNA replication; it joins RNA nucleotides to create a **primer**, which is required in order for a DNA polymerase to proceed.

▌ **Helicases** are enzymes that are responsible for unwinding the DNA helix as replication proceeds, and **single-strand binding proteins** hold the strands apart for the duration of replication.

▌ There are several different factors contributing to the accuracy of DNA replication. They are

   ▪ The specificity of base pairing
   ▪ **Mismatch repair**—in which special repair enzymes fix incorrectly paired nucleotides
   ▪ **Nucleotide excision repair**—in which incorrectly placed nucleotides are excised by an enzyme called a nuclease, and the gap left over is filled in with the correct nucleotides

■ The fact that DNA polymerase can only add nucleotides to the 3' end of a molecule would mean that it would have no way to complete the 5' end of the molecule. Thus the linear chromosomes of eukaryotes utilize an enzyme called **telomerase**, which catalyzes the ends of the molecules, called **telomeres**.

From Gene to Protein: The Connection Between Genes and Proteins

■ The **one gene-one polypeptide hypothesis** states that each gene codes for a polypeptide, which can be—or can constitute a part of—a protein.

■ **Transcription** is the synthesis of RNA using DNA as a template. It takes place in the nucleus of eukaryotic cells.

■ **Messenger RNA**, or **mRNA**, is the type of RNA produced during transcription. It carries the genetic message of DNA to the protein making machinery of the cell in the cytoplasm.

■ **Translation** is the synthesis of polypeptides. Translation takes place in the cytoplasm of eukaryotic cells at ribosomes.

■ In eukaryotes, transcription results in **pre-mRNA**, which undergoes RNA editing and processing to yield the final mRNA that participates in translation.

■ In DNA, the instructions for building a polypeptide chain are written as a series of three-nucleotide groups; this is called a **triplet code**, or codon.

■ In transcription, only one strand of the DNA is transcribed, and it is called the **template strand**. The mRNA that is produced is said to be **complementary** to the original DNA strand.

■ The complementary strand is made up of triplets called **codons** that are read, or translated, in the 3' to 5' direction along the mRNA. Each codon specifies one of the 20 amino acids, which are incorporated into a growing polypeptide strand.

■ The genetic code is **redundant**, meaning that more than one codon codes for each of the 20 amino acids. The codons are read based on a consistent **reading frame**—the groups of 3 must be read in the correct groupings in order for translation to be successful.

The Synthesis and Processing of RNA

■ The enzyme **RNA polymerase** separates the two DNA strands and connects the RNA nucleotides as they base pair along the DNA template strand.

■ The RNA polymerases can add only RNA nucleotides to the 3' end of the strand, so RNA elongates in the 5' to 3' direction.

■ The DNA sequence at which RNA polymerase attaches is called the **promoter** sequence.

■ The DNA sequence that signals the end of transcription is called the **terminator**.

■ The entire stretch of DNA that is transcribed into mRNA is called a **transcription unit**.

There are three main stages of transcription:

1. **Initiation**: In prokaryotes, a group of proteins plus RNA polymerase, bound to the promoter region of a DNA sequence, is collectively known

as a **transcription initiation complex**. In eukaryotes, the process of initiation is more complex, but it also involves the binding of RNA polymerase to a promoter sequence.

2. **Elongation**: RNA polymerase moves along the DNA, continuing to untwist the double helix. RNA nucleotides are continually added to the 3' end of the growing chain, and as the complex moves down the DNA strand, the double helix re-forms, with the new RNA molecule straggling away from the DNA template.

3. **Termination**: This occurs after RNA polymerase transcribes a terminator sequence in the DNA, and the transcribed RNA sequence is the actual termination signal.

▌ In eukaryotes, there are a couple of key post-transcription modifications to RNA—the addition of a **5' cap**, and the addition of a **poly-A tail**.

▌ Another process, called **RNA splicing**, also takes place in eukaryotic cells. In RNA splicing, large portions of the newly synthesized RNA strand are removed, or spliced out. The sections of the mRNA that are spliced out are called **introns**, and the sections that are left over—and subsequently spliced together by a **spliceosome**—are called **exons**.

The Synthesis of Protein

▌ **Translation** is the synthesis of a polypeptide, under the direction of a ribosome.

▌ **tRNA** is a type of RNA that functions in transferring particular amino acids from a pool of amino acids in the cell's cytoplasm to a ribosome. The ribosome takes the amino acid from tRNA and incorporates it into a growing polypeptide chain.

▌ Each type of tRNA is specific for a particular amino acid; at one end it loosely binds the amino acid, and at the other end it has a nucleotide triplet called an **anticodon**, which allows it to pair specifically with a complementary codon on the mRNA.

▌ The mRNA is read codon by codon, and one amino acid is added to the chain for each codon read.

▌ In translation, **wobble** refers to the fact that the third nucleotide of a tRNA can form hydrogen bonds with more than one kind of base in the third position of a codon.

▌ **Ribosomes** are made up of two subunits, one large and one small. The subunits are made up of proteins and RNA molecules called **ribosomal RNA (rRNA)**.

▌ Ribosomes have three binding sites for mRNA:

  ▪ a **P site**, which holds the tRNA that carries the growing polypeptide chain;
  ▪ an **A site**, which holds the tRNA that carries the amino acid that will be added to the chain next;
  ▪ and an **E site**, which is the exit site.

■ Translation, like transcription, can be divided into three stages:

1. **Initiation**: mRNA, a tRNA that has the first amino acid of the polypeptide, and the two ribosomal subunits come together to form a translation initiation complex. Initiation factors (proteins) are also required in order for translation to begin.

2. **Elongation**: Amino acids are added one by one to the growing polypeptide chain. Proteins called elongation factors are involved in this. Elongation involves the recognition of codons by anticodons, the formation of peptide bonds between amino acids added to the chain, and translocation—in which the tRNA in the A site is moved to the P site, and the tRNA in the P site is moved to the E site.

3. **Termination**: A stop codon in the mRNA is reached and translation stops. UAA, UAG, and UGA are all stop signal codons. A protein called release factor binds to the stop codon and the polypeptide is freed from the ribosome.

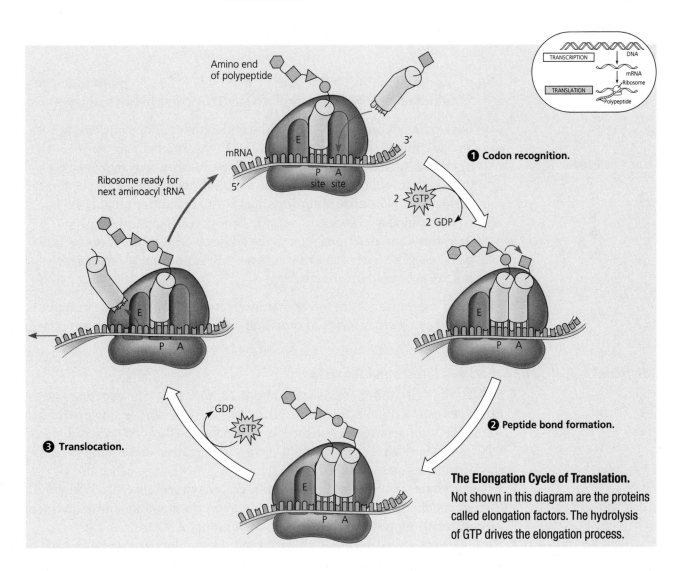

The Elongation Cycle of Translation.
Not shown in this diagram are the proteins called elongation factors. The hydrolysis of GTP drives the elongation process.

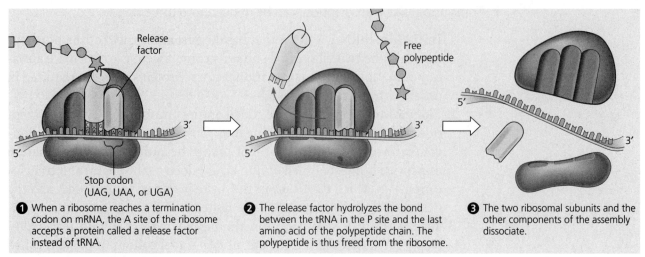

1. When a ribosome reaches a termination codon on mRNA, the A site of the ribosome accepts a protein called a release factor instead of tRNA.

2. The release factor hydrolyzes the bond between the tRNA in the P site and the last amino acid of the polypeptide chain. The polypeptide is thus freed from the ribosome.

3. The two ribosomal subunits and the other components of the assembly dissociate.

Release factor

Free polypeptide

Stop codon (UAG, UAA, or UGA)

The Termination of Translation

▌ Polypeptides then fold to assume their specific conformation, and they are sometimes postranslationally modified to render them functional.

Mutations

▌ **Mutations** are alterations in the genetic material of the cell; **point mutations** are alterations of just one base pair of a gene. They come in two basic types:

- ▪ **Base-pair substitutions** refers to the replacement of one nucleotide and its complementary base pair in DNA with another pair of nucleotides. **Missense mutations** are those substitutions that enable the codon to still code for an amino acid, although it might not be the correct one. **Nonsense mutations** are those substitutions that change a regular amino acid codon into a stop codon, ceasing translation.
- ▪ **Insertions** and **deletions** refer to the additions and losses of nucleotide pairs in a gene. They can cause a **frameshift mutation**, which cause the mRNA to be read incorrectly.

▌ Mutagens are substances or forces that interact with DNA to cause mutations. X-rays and some other forms of radiation are known mutagens, as are certain chemicals.

Microbial Models: The Genetics of Viruses and Bacteria

▌ Smaller than ribosomes, **viruses** are about 20nm across. The genetic material of viruses can be double or single-stranded DNA, or double or single-stranded RNA The viral genome is enclosed by a protein shell called a **capsid**. Some viruses also have viral envelopes that surround the capsid and aid the viruses in infecting their hosts.

▌ **Bacteriophages**, which are also called **phages**, are viruses that infect bacteria.

▌ Viruses are energy **parasites**; they can only reproduce within a host, and each virus can only reproduce within a particular group of hosts.

▌ Some viruses (called **virulent phages**) have a reproductive cycle that ends in the death of the host cell, and this is called the **lytic cycle.** In the course of the lytic cycle, the phage attaches to receptors on the cell surface, injects its DNA into the host, and directs the replication of its own DNA. New phages are assembled, and the viruses produce lysozyme, which disintegrates the cell wall so that the cell dies and new phages are released.

▌ Some viruses undergo what is called the **lysogenic** cycle, in which the host cell is not killed. In the **lysogenic cycle**, the phage DNA becomes incorporated into the host cell's DNA and is replicated along with the host cell's genome.

▌ **Retroviruses** are RNA viruses that use the enzyme **reverse transcriptase** to transcribe DNA from an RNA template. The new DNA then integrates into a chromosome in the nucleus of an animal cell, and the host transcribes the viral DNA into RNA. **HIV** is a retrovirus.

▌ **Prions** are misfolded proteins that cause the misfolding of normal proteins when they come into contact with one another. Prions are infectious proteins.

The Organization and Control of Eukaryotic Genomes: Eukaryotic Chromatin Structure and Genome Organization at the DNA Level

▌ In eukaryotic cells, DNA and proteins are packed together as **chromatin**. The first level of packing in chromosomes involves DNA and proteins called **histones**, in nearly equal amounts, folded together to resemble beads on a string. **Nucleosomes** are the basic unit of DNA packing—they are the complex of DNA and histones.

▌ In interphase, the chromosomes are extended in the nucleus (and at this point are called **euchromatin**), but as the cell prepares for mitosis, the chromosomes become packed into a form referred to as **heterochromatin**.

▌ Genes make up only a small portion of the genomes of most eukaryotes; only about 97% of the DNA does not code for proteins. These noncoding regions of DNA are made up of **introns**, **repetitive sequences**, and sequences whose function is not yet understood. Repetitive DNA is responsible for a number of genetic disorders.

▌ A group of identical or similar genes is called a **multigene family**, and it is likely that the members of the family evolved from a single ancestral gene.

▌ In **gene amplification**, certain genes are replicated into RNA in order to increase the expression of those genes; this is necessary at certain stages of development.

▌ The shuffling of sections of DNA is more common than gene amplification. **Transposons** are sections of DNA that can move from one location to another location within a genome. **Retrotransposons** can move from place to place in a genome only with the help of an RNA intermediate.

The Control of Gene Expression

▌ The expression of genes can be turned off and on at any point along the pathway from gene to functional protein.

▌ Genes in heterochromatin (which is highly packed) usually are not transcribed; this is one form of gene control. **DNA methylation** (the addition of methyl groups) is one way in which the transcription of genes is controlled.

Apparently methylation of DNA is responsible for the long-term inactivation of genes.

▌ In **histone acetylation**, acetyl groups are added to amino acids of histone proteins; this makes the chromatin less tightly packed and encourages transcription.

▌ **Transcription initiation** is another important control point in gene expression. At this stage, DNA control elements that bind transcription factors (needed to initiate transcription) are involved in regulation.

▌ Gene control also occurs after transcription, and during RNA processing, in **alternative RNA splicing**.

▌ The control of gene expression also occurs both prior to translation and just after translation, when proteins are processed.

The Molecular Biology of Cancer

▌ **Oncogenes** are cancer-causing genes; **protooncogenes** are genes that code for proteins that are responsible for normal cell growth. Protooncogenes become oncogenes when a mutation occurs that causes an increase in the product of the protooncogene—or that increases the activity of the protooncogene itself.

▌ **Cancer** can also be caused by a mutation in a gene whose products normally inhibit cell division. These genes are called tumor-suppressor genes.

▌ The incidence of cancer increases with age because multiple somatic mutations are required to produce a cancerous cell.

DNA Technology and Genomics: DNA Cloning

▌ **Recombinant DNA** is DNA that has been artificially made, using DNA from different sources and often different species. An example is the introduction of a human gene into an *E. coli* bacterium.

▌ **Genetic engineering** is the process of manipulating genes and genomes.

▌ **Biotechnology** is the process of manipulating organisms for the purpose of making products for the public.

▌ **Gene cloning** is the process by which scientists can create significant samples of specific segments of DNA that they can then manipulate in the lab.

▌ Enzymes that can be used to cut strands of DNA at specific locations (called **restriction sites**) are called **restriction enzymes**. These enzymes are what make genetic engineering possible.

▌ When a DNA molecule is cut by restriction enzymes, the result will always be a set of **restriction fragments**, which will have at least one single-stranded end, called a **sticky end**. Sticky ends can hydrogen bond with complementary single-stranded pieces of DNA, and these unions can be sealed with an enzyme called **DNA ligase**.

▌ The cloning of genes generally occurs in five steps:

1. First the vector and the gene must both be isolated. The vector is the plasmid (usually bacterial) that will carry the DNA sequence to be cloned.
2. The DNA in question must be inserted into the vector (the plasmid).
3. The vector must be inserted into the cell in order to be copied.

4. The cells must be cloned.

5. The cells carrying the clones must be identified and isolated.

▌ A **genome library** is a set of thousands of recombinant plasmid clones, each of which has a piece of one original genome being studied. A **cDNA library** is made up of complementary DNA made from mRNA transcribed from a number of different genes at one particular time.

▌ PCR (the **polymerase chain reaction**) is a method used to greatly amplify a particular piece of DNA without the use of cells.

DNA Analysis and Genomics

▌ **Genomics** is the study of an organism's entire genome.

▌ **Gel electrophoresis** is a lab technique that is used to separate macromolecules on the basis of their size and charge by using an electrical current.

▌ **Southern blotting** is a technique that is used to determine the presence of specific nucleotide sequences in DNA.

▌ The **Human Genome Project** was begun in 1990 as an effort to map the entire human genome. It was completed in 2002.

▌ The comparison of genomes of different organisms suggests strong evolutionary relationships between organisms that appear very different externally.

▌ In **proteomics**, the entire set of proteins that are encoded by genomes are studied.

▌ Bioinformatics is the application of computer science and math to genetics and other biological information.

Practical Applications of DNA Technology

There are many different uses for the work of DNA technology, some of which are:

1. **Diagnosis of diseases**—If one knows the sequence of a particular virus's DNA, one can use PCR to amplify blood samples to detect even small traces of the virus, catching it early on.

2. **Gene therapy** is the alteration of a person's genes if that person possesses problem genes of some kind. Examples of human diseases that can be treated in the future might be sickle cell anemia, Tay-Sachs disease, and Huntington's disease.

3. The creation of **pharmaceuticals**—Gene splicing and cloning can be used to create large amounts of particular proteins in the lab.

4. **Forensic applications**—DNA samples taken from the blood, skin cells, or hair of alleged criminal suspects can be compared to DNA collected from the crime scene. **DNA fingerprints** (small sets of markers, or specific bands that are unique to each individual) can be compared and used to identify persons at that crime scene.

5. DNA fingerprinting can also be used to determine **paternity**.

6. Genetic engineering is also used to help solve **environmental problems**.

7. The creation of **transgenic organisms** (organisms that carry genes from other organisms in order to enhance certain of their characteristics that humans find useful).

8. **Genetic engineering in plants**—Certain genes that produce desirable traits in plants have been inserted into crop plants to increase their productivity or efficiency.

For Additional Review

Consider the similarities and differences between prokaryotic cells and eukaryotic cells, including their structure, how they replicate their DNA, and how they live in the world.

Multiple-Choice Questions

1. A couple has 6 children, all daughters. If the woman has a seventh child, what is the probability that the seventh child will be a daughter?
   (A) $\frac{6}{7}$
   (B) $\frac{1}{7}$
   (C) $\frac{1}{36}$
   (D) $\frac{1}{49}$
   (E) $\frac{1}{2}$

2. If alleles R and S are linked, and (considering the effects of crossing over) the probability of gamete R segregating into a gamete is $\frac{1}{4}$, while the probability of allele S segregating into a gamete is $\frac{1}{2}$, what is the probability that both will segregate into the same gamete?
   (A) $\frac{1}{4} \times \frac{1}{2}$
   (B) $\frac{1}{4} \div \frac{1}{2}$
   (C) $\frac{1}{4} + \frac{1}{2}$
   (D) $\frac{1}{4} + \frac{1}{4}$
   (E) $\sqrt{\frac{1}{2}}$

3. In llamas, coat color is controlled by a gene that exists in two allelic forms. If a homozygous yellow llama is crossed with a homozygous brown llama, the offspring have gray coats. If two of the gray-coated offspring were crossed, what percentage of their offspring would have brown coats?
   (A) 100%
   (B) 75%
   (C) 50%
   (D) 25%
   (E) 0%

4. All of the following are true of the process of mitosis EXCEPT
   (A) during prometaphase, spindle microtubules come into contact with chromosomes
   (B) the chromosome number in the newly formed cells is half that of the parent cell
   (C) the chromosomes line up along the metaphase plate, or equator of the cell
   (D) the cytoplasm of the cell and all its organelles are divided approximately in half
   (E) in anaphase, the sister chromatids travel to opposite ends of the cell

5. In rabbits, the trait for short hair (S) is dominant, and the trait for long hair (s) is recessive. The trait for green eyes is dominant (G) and the trait for blue eyes is recessive (g). A cross between two rabbits produces a litter of 6 short-haired rabbits with green eyes, and 2 short-haired rabbits with blue eyes. What is the most likely genotype of the parent rabbits in this cross?
   (A) $ssgg \times ssgg$
   (B) $SSGG \times SSGG$
   (C) $SsGg \times SsGg$
   (D) $SsGg \times SSGg$
   (E) $ssGG \times ssGG$

6. In most organisms, reproduction occurs sexually; two parents are involved and genetic recombination occurs. However, some organisms reproduce asexually, and only one parent contributes genetic material. Which of the following organisms reproduces asexually?
   (A) Annelids
   (B) Arthropods
   (C) Hydra
   (D) Angiosperms
   (E) Gymnosperms

7. In humans, hemophilia is a sex-linked recessive trait. If a man and a woman produce a son that is affected with hemophilia, which of the following is definitely true?
   (A) The mother carries an allele for hemophilia.
   (B) The father carries an allele for hemophilia.
   (C) The father is affected with hemophilia.
   (D) Neither parent carries an allele for hemophilia.
   (E) The boy's paternal grandfather is a hemophiliac.

8. Which of the following explains a significantly low rate of crossing over between two genes?
   (A) They are located far apart on the same chromosome.
   (B) They are located on separate but homologous chromosomes.
   (C) The genes code for proteins that have similar functions.
   (D) The genes code for proteins that have very different functions.
   (E) The genes are located very close together on the same chromosome.

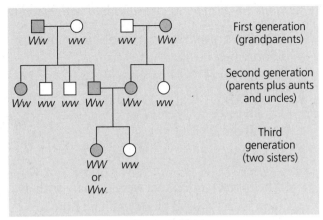

Pedigree Analysis

9. In the pedigree above, circles represent females and squares represent males; those who express a particular trait are shaded, whereas those who do not are open. Which pattern of inheritance best describes the pedigree for this trait?
   (A) Sex-linked recessive
   (B) Sex-linked dominant
   (C) Autosomal recessive
   (D) Autosomal dominant
   (E) Codominant

Questions 10–11 refer to an individual with blood type O, whose mother has blood type A.

10. The father must have which of the following blood types?
    (A) A, B, or O
    (B) AB or A
    (C) AB or B
    (D) AB only
    (E) O only

11. If the type O individual were to mate with a person with type AB blood, which of the following is the best calculation of the ratio and genotype of the offspring?
    (A) 3:1, $I^A i$: $I^B i$
    (B) 2:1, $I^A i$: $I^B i$
    (C) 1:1, $I^A i$: $I^B i$
    (D) 1:2:1, $I^A i$: $I^A I^B$: $I^B i$
    (E) 9:3:3:1, $I^A I^B$: $I^A i$: $I^B i$: O

12. Two yellow mice with the genotype $Yy \times Yy$ are mated. Two-thirds of their offspring are yellow, and ⅓ of their offspring are not yellow (a 2:1 ratio). Mendelian genetics dictates that this cross should produce offspring that were ¼ YY (and yellow), ½ Yy (and yellow) and ¼ yy (and not yellow). What is the most likely conclusion from this experiment?
    (A) The mice did not bear enough offspring for the ratio calculation to be specific.
    (B) Y is a lethal allele and caused death early in development.
    (C) Nondisjunction occurred.
    (D) A mutation masked the effects of the Y allele.
    (E) A mutation masked the effects of the y allele.

13. In organisms that undergo alternation of generations, the diploid stage is called the
    (A) gametophyte
    (B) ovum
    (C) sporophyte
    (D) hybrid
    (E) dicot

14. All of the following contribute to genetic recombination EXCEPT
    (A) random fertilization
    (B) independent assortment
    (C) crossing over
    (D) gene linkage
    (E) random gene mutation

15. In cucumbers, warty (W) is dominant over dull (w), and green (G) is dominant over orange (g). A cucumber plant that is homozygous for wartiness and green color is crossed with one that is homozygous for dullness and orange color. The F_1 are then crossed to produce an F_2 generation. If a total of 144 offspring are produced, which of the following is the closest to the number of dull green cucumbers expected?
    (A) 3
    (B) 10
    (C) 28
    (D) 80
    (E) 161

16. Which of the following exists as DNA surrounded by a protein coat?
    (A) A retrovirus
    (B) A virus
    (C) A eukaryote
    (D) A prokaryote
    (E) Ampicillin

17. A goat can be made to produce milk containing the same polymers present in the silk produced by spiders when particular genes from a spider are inserted into the goat's genome. Which of the following reasons describes why this is possible?
    (A) Goats and spiders share a common ancestor and thus produce similar protein excretions.
    (B) The opposite is true, too—when genes from a goat are inserted into a spider's genome, the spider produces goats' milk instead of silk.
    (C) The proteins in goats' milk and spiders' silk have the same amino acid sequence.
    (D) The processes of transcription and translation in the cells of spiders and goats are fundamentally similar.
    (E) The processes of transcription and translation in the cells of spiders and goats produce exactly the same proteins anyway.

Directions: The group of questions below consists of five lettered headings followed by a list of numbered phrases or sentences. For each numbered phrase or sentence select the one heading that is most closely related to it and fill in the corresponding oval on the answer sheet. Each heading may be used once, more than once, or not at all.

Questions 18–22
    (A) Transcription
    (B) Translation
    (C) Transposon
    (D) DNA methylation
    (E) Histone acetylaction

18. A mobile segment of DNA that travels from one location on a chromosome to another, one element of genetic change

19. The addition of methyl groups to certain bases of DNA after DNA synthesis, this is thought to be an important control mechanism for gene expression.

20. The synthesis of polypeptides from the genetic information coded in mRNA

21. The synthesis of RNA from a DNA template

22. The attachment of acetyl groups to particular amino acids of histone proteins, this is thought to be an important control mechanism for gene expression.

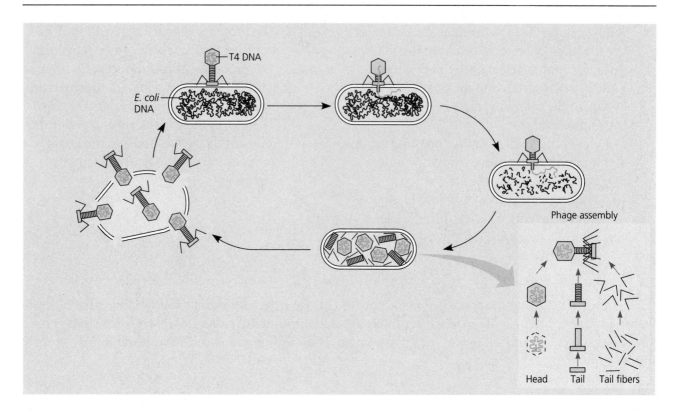

23. The above figure shows which of the following processes?
    (A) The lytic cycle of a phage
    (B) The lysogenic cycle of a phage
    (C) Transcription
    (D) Translation
    (E) DNA Replication

24. Restriction enzymes are generally used in the laboratory for which of the following reasons?
    (A) Restricting the replication of DNA
    (B) Restricting the transcription of DNA
    (C) Restricting the translation of mRNA
    (D) Cutting DNA molecules at specific locations
    (E) Cutting DNA into manageable sizes for manipulation

Questions 25–28 refer to an experiment that was performed to separate DNA fragments from 3 samples radioactively labeled with ^{32}P. The fragments were then separated using gel electrophoresis. The visualized bands are depicted below:

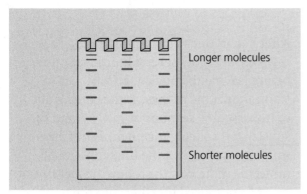

Gel Electrophoresis: Completed Gel

25. When the electric field was applied, the fragments of DNA in each of the 3 samples migrated to different locations along the gel because
    (A) the fragments differed in their levels of radioactivity
    (B) the fragments differed in their charges—some were positively charged while others were negatively charged
    (C) the fragments differed in size
    (D) the fragments differed in polarity
    (E) the fragments differed in solubility

26. How many sites on DNA were cut by the particular restriction enzyme used in Sample 1 (the left-most sample)?
    (A) 5
    (B) 6
    (C) 7
    (D) 8
    (E) 9

27. The DNA in this experiment was labeled with ^{32}P because
    (A) without ^{32}P, the DNA would not migrate through the gel
    (B) without ^{32}P, we would not be able to visualize the DNA fragments
    (C) ^{32}P is required in order to allow the restriction enzymes to make their cuts
    (D) radioactivity limits the interference of scrap fragments of DNA
    (E) radioactivity enables the DNA fragments to clump together and produce bands

28. Gel electrophoresis can also be used for which of the following purposes?
    (A) To group molecules based on their polarity
    (B) To measure the acidity of certain large molecules
    (C) To measure the polarity of certain large molecules
    (D) To separate out the proteins in a mixture
    (E) To measure the amount of protein in a mixture of substances

29. In genetic engineering, DNA ligase is used for which of the following purposes?
    (A) To act as a probe for locating cloned genes
    (B) To create breaks in DNA in order to allow foreign DNA fragments to be inserted
    (C) To seal up nicks created in newly created recombinant DNA
    (D) To ensure that "sticky ends" of like DNA fragments do not re-anneal
    (E) In Southern blotting

The group of questions below consists of five lettered headings followed by a list of numbered phrases or sentences. For each numbered phrase or sentence select the one heading that is most closely related to it and fill in the corresponding oval on the answer sheet. Each heading may be used once, more than once, or not at all.

Questions 30–34
  (A) tRNA
  (B) mRNA
  (C) poly-A tail
  (D) RNA polymerase
  (E) rRNA

30. An example of a post-transcriptional modification

31. Binds to the promoter on DNA to initiate transcription

32. Along with proteins, comprise ribosomes

33. Loosely binds to free amino acids in the cytoplasm

34. Travels out of the nucleus and into the cytoplasm to participate in translation

35. All of the following nitrogenous bases are included in DNA EXCEPT
  (A) adenine
  (B) cytosine
  (C) guanine
  (D) thymine
  (E) uracil

36. The actions of which of the following enzymes are responsible for ensuring that chromosomes do not decrease in length with every round of replication?
  (A) Telomerase
  (B) DNA ligase
  (C) DNA polymerase
  (D) Helicase
  (E) Primase

37. After eukaryotic transcription takes place, mRNA undergoes several modifications before leaving the nucleus to take part in translation. One of these is the cutting out of nonessential sections of mRNA, and the subsequent splicing together of stretches of mRNA necessary for the final functional molecule. Which of the following sections of mRNA are spliced together into the finished mRNA molecule?
  (A) Introns
  (B) Exons
  (C) Genes
  (D) Coding sequences
  (E) Ribozymes

38. In the process of eukaryotic translation, the term *wobble* refers to
  (A) the tendency of the two ribosome subunits to come closer to one another and to separate at different points in translation
  (B) the tendency of the amino acid loosely attached to the tRNA to move back and forth before finally attaching to the polypeptide chain
  (C) the fact that the genetic code is redundant
  (D) the fact that the anticodon and codon bind very loosely
  (E) the fact that the third nucleotide of a tRNA can form hydrogen bonds with more than one kind of base in the third position of a codon

39. Which of the following is an example of a missense mutation?
    (A) A nucleotide and its partner are replaced with an "incorrect" pair of nucleotides, which destroys the function of the final protein.
    (B) A nucleotide pair is added into a gene, destroying the reading frame of the genetic message.
    (C) A nucleotide pair is lost from the gene, destroying the reading frame of the genetic message.
    (D) A frameshift mutation occurs, ultimately causing the production of a nonfunctional proteins.
    (E) A nucleotide pair substitution occurs, which causes the codon to code for an amino acid that may not be the "correct" one, although translation continues.

40. Which of the following is an example of a nonsense mutation?
    (A) A chemical change occurs in just one base pair of a gene, and it has no effect on the final protein.
    (B) Part of the gene breaks off and travels to a distant location on the chromosome, inserting itself there.
    (C) A substitution occurs which changes a regular amino acid codon into a stop codon, causing translation to cease.
    (D) A substitution occurs, which changes a regular amino acid codon into a start codon, and translation begins again, creating two unfinished polypeptides.
    (E) A nucleotide pair substitution occurs, which causes the codon to code for an amino acid that may not be the "correct" one, although translation continues.

41. PCR—the polymerase chain reaction—makes gene cloning possible because it enables us to do which of the following very quickly?
    (A) Isolate gene-source DNA
    (B) Insert DNA into an appropriate vector
    (C) Introduce the cloning vector into a host cell
    (D) Amplify DNA samples
    (E) Identify clones carrying the gene of interest

42. At the end of DNA replication, each of the daughter molecules has one old strand, derived from the parent strand of DNA, and one strand that is newly synthesized. This explains why DNA replication is described as
    (A) conservative
    (B) largely conservative
    (C) nonconservative
    (D) semiconservative
    (E) unconservative

43. The expression of genes can be controlled at all the following stages of protein synthesis EXCEPT
    (A) initiation of transcription
    (B) RNA processing
    (C) DNA unpacking
    (D) degradation of protein
    (E) protein folding

1. *Genes are located on chromosomes and are the basic unit of heredity that are passed on from parent to child, through generations.*

❚ Explain how a chromosome mutation could occur, and why mutations are detrimental to the organism in which they take place.

❚ Explain why it is that—although there are very few genes located on the Y chromosome—human males suffer from having just one copy of the X chromosome, while females have two.

ANSWERS AND EXPLANATIONS

Multiple-Choice Questions

❚ **1. (E) is correct.** The probability that the woman will have a seventh child who is a daughter is ½. Since the probability that a sperm carrying an X chromosome and the probability that a sperm carrying a Y chromosome will fertilize an egg is equal—both 50%—fertilization is considered an independent event. The outcome of independent events is unaffected by what events occurred before or will occur after. Therefore, the probability that this woman's next child will be a girl is ½. Likewise, the probability that she will have a child that is a boy is also 50%.

❚ **2. (A) is correct.** If the probability of allele R segregating into a gamete is ¼, and that of S segregating is ½, then *how* they segregate is not relevant. Rather, you can calculate the probability of two independent events occurring in a specific combination, order, or sequence by multiplying their probabilities. So in this case, you need to multiply ¼ × ½.

❚ **3. (D) is correct.** The question tells you that one parent is homozygous for yellow coat, so you can use any designation for that—*BB* or *bb*. (It hasn't specified which is dominant and which is recessive, but either way it doesn't matter.) Let's say that the yellow coat parent is *BB*, and the homozygous brown coat parent is *bb*. Since the yellow coat parent can only produce gametes *B*, and the brown coat parent can only produce gametes *b*, the F_1 generation will all have genotype *Bb*. Crossing two members of this generation would give you a ratio of 1 yellow coat: 2 gray coats: 1 brown coat. This means that 25% of the offspring would have brown coats, 25% would have yellow coats, and 50% would have gray coats.

❚ **4. (B) is correct.** All of the statements about mitosis are true except the one about the daughter cells being haploid while the parent cells are diploid. This is true of meiosis, which occurs in the gamete-producing cells of the gonads (testes and ovaries) but not of mitosis, which takes place in the somatic cells.

❚ **5. (D) is correct.** To find the answer to this problem, first look at the ratio of the offspring. It's 6:2, which can be reduced to 3:1. Next, you can quickly work through the crosses listed. You can immediately rule out answers A and B, because A would give you only offspring that exhibited the dominant traits, short hair and green eyes, and B would give you all offspring that had the recessive traits—long hair and blue eyes. If you look carefully at the remaining answers, you will want to choose the one that will give you all short-haired off-spring, so you will need the dominant allele to be present in both parents. This rules out answer *E*. If you still cannot choose between *C* and *D*, write out what gametes the parents could produce, and then use a Punnett square to determine their offspring. By doing this, you can see that *D* is correct: the ratio of offspring is 12:4, or 3:1, which matches the ratio in the original question.

■ **6. (C) is correct.** The only organism listed in the answer choices that reproduces sexually is Hydra. Hydra undergo a form of reproduction called budding, in which a mass of mitotically dividing cells "buds" off from the parent, eventually developing into a small hydra. In the process, all of its copies of its genes are passed on to the offspring.

■ **7. (A) is correct.** If the boy is affected with hemophilia, then he must have inherited the recessive hemophilia gene from his mother. Sex-linked genes are usually located on the X chromosome. In order for the child to be a boy, he must have inherited the Y chromosome from his father. Since the gene causing hemophilia is located on the X chromosome, you can rule out answers *B* and *C*, (because it would not matter if the father possessed the allele for hemophilia or even had the disease, since he can't pass on his X to a son). Therefore, you need to look for an answer choice that shows that the source of his X chromosome was a carrier of the allele—affected or not. This is answer *A*.

■ **8. (E) is correct.** While genes that are on the same chromosome tend to be inherited together, the process of crossing over enables "linked" genes to assort independently. Those that are linked but located farther apart on the chromosome will undergo crossing over more frequently than those located very close together on a chromosome, simply because there are more sites between the two genes at which crossing over can take place.

■ **9. (D) is correct.** Autosomal dominant traits appear with equal frequency in both sexes, and they do not skip generations. These qualities are all exhibited by the trait that is illustrated in the pedigree. All three generations are affected with the trait; the sexes are affected roughly equally (4 women and 2 men are affected).

■ **10. (A) is correct.** The father either has type A, B, or O blood. The mother, who has the phenotype blood type A, has the genotype $I^A i$. In order to produce a son with genotype *ii* (the genotype of people with blood type O), she would need to reproduce with a man who had genotype $I^A i$, genotype $I^B I$, or genotype *ii*. Try writing out the Punnett square for yourself, if you aren't confident of this.

■ **11. (C) is correct.** If the type O individual were to mate with an individual who was type AB, since I^A and I^B are both dominant over I, the genotype would be 1:1 I^A:I^B, and the phenotype would be a ratio of 1:1 offspring with type A or type B blood.

■ **12. (B) is correct.** The most likely reason for this 2:1 ratio in the offspring is that *Y* is lethal in homozygous form, and this caused the death of all of the *YY* individuals in the litter. If you look at the phenotypes of the offspring considering that this occurred, you get 1*YY*: 2*Yy*: 1*yy*, so minus the 1*YY*, you get a ratio of 2*Yy* (yellow mice, since the gene for yellow, *Y*, is dominant) to 1*yy* (non-yellow mouse).

■ **13. (C) is correct.** Alternation of generations occurs in plants and some algae, and organisms that undergo alternation of generations have both a haploid and diploid stage in their life cycle. The diploid stage is the sporophyte, and meiosis in the diploid phase creates haploid spores, which divide mitotically to produce a gametophyte. The gametophyte produces haploid gametes through mitosis, and the fertilization occurs to produce a diploid zygote.

■ **14. (D) is correct.** The only process listed that does *not* lead directly to genetic recombination, or the recombining (scrambling) of genes in the offspring, is gene linkage. If genes are linked they are located on the same chromosome, and are more likely to segregate together into the same cell.

■ **15. (C) is correct.** If you haven't memorized the fact that a dihybrid cross between two heterozygotes produces a 9:3:3:1 ratio—or noticed that the question is asking for one of the heterozygotes (which would be one of the 3's in the above ratio)—then you could complete the Punnett square and see that the ratio of offspring produced is 9 warty green, 3 warty orange, 3 dull green, and 1 dull orange. If the total number of offspring produced is 144, then you can deduce that that's 9 times 9 + 3 + 3 + 1, which equals 16. So the number of warty green cucumbers must be equal to $9 \times 3 = 27$, and the closest answer choice to 27 is 28.

■ **16. (B) is correct.** Viruses are made up of nucleic acid (DNA in the case of regular viruses, RNA in the case of retroviruses) surrounded by a protein coat. They reproduce by injecting their genetic material into a host cell, and using the cell's replicative machinery to replicate their DNA and proteins, they self-assemble and leave the cell, sometimes killing it in the process.

■ **17. (D) is correct.** The process of genetic engineering is possible because of the fact that, in all eukaryotic cells, transcription and translation occur in similar ways. Once the spider gene or genes that were responsible for coding for the silk proteins were isolated and then inserted into a bacterial plasmid (which would serve as the vector), the cloning vector would be taken up by the goat's cells, and the goat's cells transcription/translation machinery would begin the process of producing the spider protein, along with its own proteins.

■ **18. (C) is correct.** Transposons are also called transposable genetic elements, and they are pieces of DNA that can move from location to location in a chromosome—or a genome. Transposons are also called "jumping genes," and most of them are capable of moving to many different target sites in the genome.

■ **19. (D) is correct.** One of the two important ways that the cell has of controlling gene expression is through DNA methylation. In DNA methylation, methyl groups are attached to certain DNA bases after DNA is synthesized. This appears to be responsible for the long-term inactivation of genes.

■ **20. (B) is correct.** The process by which genetic information flows from mRNA to protein is called translation. Translation occurs in the cytoplasm of the cell, at ribosomes. A molecule of mRNA is moved through the ribosome, and codons are translated into amino acids one at a time. tRNAs add their associated amino acids onto a growing polypeptide as its anticodon pairs with a codon on the mRNA, and then departs from the ribosome to bind more free amino acids.

■ **21. (A) is correct.** In transcription, RNA is synthesized using the genetic information encoded by DNA. Transcription occurs in the nucleus of the cell. The double-stranded DNA helix unwinds to allow enzymes and proteins to synthesize a new complementary single-stranded mRNA molecule from the parent strand of DNA.

22. (E) is correct. In histone acetylation, acetyl groups are attached to certain amino acids of histones. Deacetylation is the process by which they are removed. Acetylation makes the histones change shape so that they are less-tightly bound to DNA, and this allows the proteins involved in transcription to move in and begin the process. Therefore acetylation is one way for the cell to instigate transcription and to control the expression of its genes.

23. (A) is correct. This art portrays the lytic cycle of phage reproduction. In the lytic cycle, the phage first attaches to the cell surface and injects its DNA into the cell. It then hydrolyzes the host cell's DNA and uses the cell's machinery to produce phage proteins and to replicate its genome. The phage proteins are then assembled in the cell and produce lysozyme, which breaks down the cell wall, so that the phage can exit. In the lysogenic cycle, the phage genome becomes incorporated into the host cell's DNA without destroying the host cell.

24. (D) is correct. Restriction enzymes can be used to cut DNA at specific locations, and this enables us to perform recombinant DNA techniques. When specific restriction enzymes are added to the DNA, they produce cuts in the sugar-phosphate backbone and create "sticky ends," which can bind to DNA fragments from a different source to produce recombinant DNA. DNA ligase is then added to seal the strands together permanently.

25. (C) is correct. The fragments of DNA separated out from one another along the gel once the electric field was applied because they differed in size. For DNA in gel electrophoresis, how far a molecule travels along a gel (while the current is applied) will be inversely proportional to its size. This means that the larger a fragment is, the more slowly it will migrate.

26. (D) is correct. The restriction enzyme used to cut the DNA that was placed into the first well of the gel must have cut the DNA at 8 sites, because it produced 9 DNA fragments. The number of fragments produced is always one more than the number of restriction sites cut.

27. (B) is correct. Radioactivity is conferred to the DNA fragments so that we can visualize them once they have ceased migrating through the gel. This can be done by applying a piece of film to the gel—the radioactivity exposes the film to form an image that corresponds to the bands of DNA shown.

28. (D) is correct. Gel electrophoresis is a method used to separate macromolecules (DNA, protein—most types of macromolecule) based on their rate of movement through a gel once an electric field has been applied. The rate of their movement will be inversely proportional to their size.

29. (C) is correct. In genetic engineering (the manipulation of genes for practical purposes), DNA ligase is an enzyme that is used to seal the strands of newly recombinant DNA (DNA that is spliced together from two different sources) by catalyzing the formation of phosophodiester bonds.

30. (C) is correct. The addition of a poly-A tail after transcription is one example of post-transcriptional modifications that the mRNA undergoes. This poly(A) tail inhibits the degradation of the newly synthesized mRNA strand, and is thought to also help ribosomes attach to it. Another important modification that mRNA undergoes is the addition of a 5'cap. The 5' cap helps protect mRNA from degradation and also acts as the point of attachment for the ribosomes, just prior to translation.

31. (D) is correct. RNA polymerase is the most prominent enzyme involved in the transcription of DNA to create mRNA. It is responsible for binding to the promoter sequence on the parent DNA, prying the two DNA strands apart, and hooking the RNA nucleotides together as they base-pair along the DNA template. RNA polymerases add nucleotides to the 3' end of the growing chain until a terminator sequence is reached—it transcribes entire transcription units.

32. (E) is correct. Ribosomal RNA (rRNA), together with proteins, makes up ribosomes. Ribosomes, the sites of protein synthesis, are composed of two subunits, the large and the small subunit. The large subunit of the ribosome contains the A, P, and E sites, which shuttle through the tRNA and mRNA during translation.

33. (A) is correct. tRNA, or transfer RNA, interprets the genetic message coded in mRNA. It transfers amino acids taken from the cytoplasmic pool to a ribosome, which adds the specific amino acid brought to it by tRNA to the end of a growing polypeptide chain. Each type of tRNA binds loosely to a specific amino acid at one end; its other end contains an anticodon, which base-pairs with a complementary codon on the mRNA strand.

34. (B) is correct. mRNA, also known as messenger RNA, is a type of RNA that is synthesized from DNA and attaches to ribosomes in the cytoplasm to specify the primary structure of a protein. Since mRNA is the product of transcription, which occurs in the nucleus, it must travel out of the nucleus and into the cytoplasm in order to participate in translation.

35. (E) is correct. The base uracil is found in RNA but not in DNA. The bases in DNA are cytosine, guanine, thymine, and adenine, whereas the bases found in RNA are cytosone, guanine, thymine, and uracil. In DNA, cytosine is capable of forming three hydrogen bonds with guanine, and thymine and adenine form two hydrogen bonds between them—the bases form the "rungs" of the double helix ladder, and sugar-phosphate groups form the rails of the ladder.

36. (A) is correct. The enzyme responsible for adding nucleotides to the replicating DNA strand, DNA polymerase can only add nucleotides to the 3' end of a molecule. This means that it would have no way to complete the 5' end of the molecule; thus, the linear chromosomes of eukaryotes utilize an enzyme called telomerase, which catalyzes the ends of the molecules (called telomeres).

37. (B) is correct. In the modification of mRNA that occurs after transcription, a process called RNA splicing occurs. In this process, noncoding regions of nucleic acid that are situated between coding regions are cut out. These noncoding regions are called introns. The remaining regions are called exons, and these are spliced together to form the final mRNA product. When you think of exons, think *expressed*—because they are actually translated into proteins, whereas introns are not.

38. (E) is correct. In eukaryotic translation, the term *wobble* refers to the fact that more than one tRNA exists for every mRNA codon that specifies for an amino acid. Some tRNAs have codons that can recognize two or more different codons because of wobble—wherein the base-pairing rules are relaxed, and the third nucleotide of a tRNA can form hydrogen bonds with more than one kind of base in the third position of a codon.

39. (E) is correct. A missense mutation is a base-pair substitution (the replacement of a nucleotide and its partner in the cDNA strand with a different pair of nucleotides) that still enables the codon to code for an amino acid. The amino acid may or may not ultimately contribute to a functional protein, but a missense mutation is one where an amino acid is still chosen to be added to the polypeptide chain, and translation will continue.

40. (C) is correct. In a nonsense mutation, a base-pair substitution takes place—one base pair is replace by another—and the point mutation changes the codon for a regular amino acid into a stop codon. This makes translation end prematurely, and this results in a shortened and usually nonfunctional protein.

41. (D) is correct. PCR, the polymerase chain reaction, is a technique by which any piece of DNA can be copied many times without the use of cells. The DNA is heated to separate its strands, then cooled to allow primers to attach to the single strands; the DNA polymerase is added, it begins to add nucleotides to the 3' end of each primer on the two strands, and this creates two strands. With each turn of the cycle, the amount of DNA is multiplied by two.

42. (D) is correct. DNA replication is semi-conservative. Each new daughter molecule created contains one newly synthesized strand, and one strand that used to belong to the parent double helix DNA.

43. (E) is correct. Protein folding is the mechanism by which the polypeptide assumes its functional conformation. It is the only answer listed that does not describe a stage in the pathway from gene to protein that is involved in controlling gene expression. At almost all of the stages in this pathway, the cell has some mechanism for controlling the expression of its genes or the amount of gene product produced.

Free-Response Question

1. The reason that it is detrimental to an organism to have an abnormal chromosome number is that genes, which are located on chromosomes, code for proteins, which have specific functions in the cell. If an organism has two copies of a particular gene, then this gene will be transcribed twice, creating twice the usual gene product. This alters the relative amounts, or doses, of interacting products in the cell, and this can cause serious developmental problems. Likewise, if a gene is missing from a chromosome, it will not be transcribed and its corresponding protein will not be produced. If that protein has an important cellular function, the organism will be seriously affected.

There are many ways by which chromosomes can be altered to cause problems for the cell. Among these are nondisjunction—when during mitosis or meiosis the chromosomes fail to separate properly, and one cell ends up with two copies of a chromosome while the other gets no copies. This results in a condition called aneuploidy. Smaller chromosomal mutations are: deletions (in which part of the chromosome breaks off and is lost), inversions (in which a chromosome segment is reversed within a chromosome), duplications (in which a chromosome segment is repeated in a chromosome), or translocation (in which part of a chromosome is moved from one chromosome to another).

There are not very many important genes located on the Y chromosome, whereas there are many crucial sex-linked genes located on the X chromosome. However, when fertilization occurs and a sperm carrying a Y chromosome penetrates the egg first, a male zygote with one X and one Y chromosome is produced. If a sperm carrying an X chromosome penetrates the egg first, a female zygote with XX is produced. Although it seems like the female zygote would have the advantage of having twice the cell product as the male, due to its double dose of genes located on the X, this is not true. The reason for this is that, in every cell of the human female body, one of the X chromosomes is inactivated. It isn't clear yet how this happens, but the X chromosome that is inactivated condenses into a structure called a Barr body, which then associates with the nuclear envelope. As a Barr body, most of the X chromosome's genes are not expressed—although some of them do remain active. As a result of this, females are a mosaic consisting of roughly half cells that have the X chromosome from the mother inactivated, and half cells that have the X chromosome from the father inactivated. This is also the reason why sex-linked disorders are usually not expressed in females. Though one of the X chromosomes may be incapable of producing a crucial gene product, this mosaic effect insures that the other half of the cells in the body produce sufficient amounts of the protein in question.

This response demonstrates knowledge of the following terms and processes:

chromosome	*inversion*
gene	*duplication*
gene product	*translocation*
doses	*X and Y chromosome*
transcription	*zygote*
nondisjunction	*X chromosome inactivation*
aneuploidy	*Barr body*
deletion	

Moreover, the response describes the following processes in a clear, concise, and organized way:

—the function of genes and why their loss or duplication affects the cell
—the many types of chromosome mutations that can occur
—the difference between X and Y chromosomes
—the process and result of X chromosome inactivation

CHAPTER 4

Mechanisms of Evolution

Descent with Modification: A Darwinian View of Life: The Historical Context for Evolutionary Theory

▮ The concept of **natural selection** states that a population can change over the course of time if individuals with certain heritable traits produce more viable offspring than the other individuals do. The result of natural selection is **evolutionary adaptation,** which is the tendency of the characteristics of a species to change over generations to enable that species to better fit the environment.

▮ **Taxonomy** is the sector of biology dedicated to the naming and classification of all forms of life. Species are named using a two-part system (binomial nomenclature).

▮ **Fossils,** which are found in sedimentary rock, are impressions of organisms that are no longer living. Fossils have given proof of the theory of evolution. **Paleontology** is the study of fossils.

▮ **Gradualism** is a geologic theory that states that profound changes in earth's features over the course of geologic time is the result of slow, continuous processes. **Uniformitarianism** is the idea that the geologic processes that have shaped the planet have not changed over the course of the earth's history.

▮ **Jean Baptiste Lamarck** published an early theory of evolution, which stated that characteristics acquired during an organism's lifetime could be passed on to the next generation. This theory was later proven to be incorrect.

The Darwinian Revolution

▮ **Charles Darwin**'s voyage on the HMS Beagle in 1831 was the impetus for the development of his theory of evolution.

▮ The phrase **descent with modification** refers to Darwin's idea that all living organisms are related by descent (i.e., they evolved) from an unknown common ancestor in the past.

▮ **Darwin's theories** can be stated in three parts:

1. Natural selection is the result of differing reproductive success that is due to the unequal ability of individuals to survive to reproduce.
2. The process of natural selection occurs through interactions between the environment and organisms that vary in their genotype in a population.
3. The result of natural selection is the adaptation of a population to its environment.

▮ **Artificial selection** is the process by which species are modified by humans. Plants and animals are specifically chosen to breed with the desired goal of producing offspring with specific characteristics.

- A **population** is defined as a group of interbreeding individuals that live in a certain geographic area—it is the smallest unit that can evolve. For example, individuals can not evolve.
- Natural selection can work only on **heritable traits**, not on traits that can not be inherited.
- Since organisms are related evolutionarily, species that share common ancestry should have similarities. This phenomenon, in which related species share characteristics, is known as **homology**.
- **Homologous structures** are anatomical signs of evolution—one example of this is the forelimbs of mammals that are used for different purposes but have been evolved from a common ancestor.
- **Vestigial organs** are those that are historical remnants of structures that were functional in ancestors.
- Two other types of homologies are **embryological homologies** (which are most prominent during development) and **molecular homologies** (which occur when organisms share characteristics on the molecular level, such as using the same method for reproducing DNA or other cellular processes).
- **Biogeography** refers to the geographic distribution of species. Species that live closer to one another tend to be more closely related than those who do not. Species that are **endemic** to a certain geographic location are found at that location and nowhere else.

The Evolution of Populations: Population Genetics

- Natural selection acts on individuals—because their relative fitness determines if they will survive to reproduce—but evolution acts on populations.
- **Population genetics** is the study of the genetic variation that exists within populations and of how it changes.
- A **species** is a group of populations of individuals that can interbreed successfully and produce a viable offspring. Populations of the same species may be geographically isolated and only exchange genetic material rarely.
- The **gene pool** is the total aggregate of genes in a population at any one time. It is made up of all the alleles at all loci in all of the members of a population. Remember that in diploid species, each individual has two alleles, and it may be either heterozygous or homozygous.
- If all members of a population are homozygous for the same copy of an allele, the allele is said to be **fixed**.
- The **Hardy-Weinberg theory** is used to describe a population that is not evolving. It states that the frequencies of alleles and genes in a population's gene pool will remain constant over the course of generations unless they are acted upon by forces other than Mendelian segregation and the recombination of alleles. The situation in which the allele frequencies within a population are not changing is termed **Hardy-Weinberg equilibrium**.
- For a gene locus that exists in two allelic forms in a population, one having a frequency of p and one having a frequency of q, if we know the frequency of one of the alleles, we can calculate the frequency of the other allele:

$$p + q = 1, \text{so}$$
$$p = 1 - q$$
$$q = 1 - p$$

And the two are related by the **Hardy Weinberg equation**:

$$p^2 + 2pq + q^2 = 1$$

Where p^2 is equal to the frequency of the homozygous dominant in the population, $2pq$ is equal to the frequency of all of the heterozygotes in the population, and q^2 is equal to the frequency of the homozygous recessive in the population.

▌ The Hardy-Weinberg equation can be used to determine or predict the allelic frequencies that exist in populations. In order for a population to be in Hardy-Weinberg equilibrium, it must meet all of these criteria:

1. It must be a very large population size. If a population is small, a change in the gene pool due to chance will have an inordinate effect on the gene frequencies of a population.
2. There must be no migration in a population. Gene flow, which is the transfer of alleles between populations, cannot occur, because that would alter gene frequencies.
3. There can be no mutations. Mutations can change one allele into another, thus altering the gene pool.
4. Mating must occur randomly. If individuals mate only with other individuals of a certain genotype, this does not meet the criteria of a random mixing of genes.
5. Natural selection may not be taking place. This would alter gene frequencies and cause a deviation from Hardy-Weinberg equilibrium.

Causes of Microevolution

▌ Populations that deviate from the Hardy-Weinberg equilibrium are said to be **evolving**.

▌ **Microevolution** can be defined as a generation-to-generation change in allelic frequency in a population. The main causes of microevolution are genetic drift, natural selection, gene flow, and mutation. **Genetic drift** refers to a change in a population's allele frequencies due to chance.

▌ **The bottleneck effect** occurs when natural disaster or some other event causes a drastic reduction in the size of a population, which in turn causes genetic drift. Bottlenecking usually reduces the genetic variability in a population, since some alleles are lost from the gene pool.

▌ **The founder effect** occurs when a few members of a population colonize an isolated location—the smaller the number of founders, the more limited the variability of the genes in the population.

▌ Natural selection refers to the differing reproductive success of individuals in a population. The individuals best suited to their environment will survive to reproduce and to pass on their alleles to the next generation.

- **Gene flow** refers to genetic exchange due to the migration of individuals or gametes between populations.
- **Mutation** refers to a change in an organism's DNA. This can alter the gene pool of a population by changing one allele into another.

Genetic Variation, the Substrate for Natural Selection

- **Genetic variation** exists naturally in populations. Quantitative characters, such as height in humans, vary in a continuum in a population.
- A population is said to be **polymorphic** for a character if this character exists in two or more discrete forms in the population—for example, if a plant bears two different kinds of flowers in a population. **Geographic variation** refers to differences in gene pools between populations or parts of populations. A **cline** is a graded change in a trait along a geographic axis.
- Two processes contribute to variation in the gene pool of a population, one is **mutation** and the other is **sexual recombination**.
- Since mutations in somatic cells disappear when the individual dies, only mutations in gametes are passed to offspring.
- Most of the genetic differences that exist in a population are due to the genetic recombination of alleles that already exist in a population.
- Factors that contribute to the preservation of genetic variation in a population are **diploidy** and **balanced polymorphism**. The fact that most eukaryotes are diploid means that they are capable of hiding genetic variation (recessive alleles) from selection. Balanced polymorphism refers to the ability of natural selection to keep stable the frequencies of two or more phenotypes in a population.
- Individuals with **heterozygote advantage** are heterozygous at a certain locus, and this confers upon them an advantage that enables them to better survive. An example of this is seen in the case of sickle-cell anemia.

The Origin of Species: What Is a Species?

- **Macroevolution** refers to the origin of new taxonomic groups.
- **Speciation** is the process by which a new species arise.
- There are many barriers that prevent members of different species from interbreeding; these barriers can be broken down into two types—**prezygotic barriers** (those that prevent mating between species or hinder fertilization) and **postzygotic barriers** (those that prevent a fertilized egg from developing into a fertile adult). Examples of prezygotic barriers include:

  1. **Habitat isolation**—Two species can live in the same geographic area, but not in the same habitat; this will prevent them from mating.
  2. **Behavioral isolation**—Some species use certain signals or types of behavior to attract mates, and these signals are unique to their species. Members of other species will not recognize them, which prevents them from mating.
  3. **Temporal isolation**—Species may breed at different times of the day, different seasons, or different years, and this can prevent them from mating.
  4. **Mechanical isolation**—Species may be anatomically incompatible.

5. **Gametic isolation**—Even if the gametes of two species do meet, they might be unable to fuse to form a zygote.

Postzygotic barriers include:

1. **Reduced hybrid vitality**—When a zygote *is* formed, the fact that the two species are genetically incompatible may cause development to cease.
2. **Reduced hybrid fertility**—Even if the two different species produce a viable offspring, reproductive isolation is still occurring if the offspring are sterile and can't reproduce.
3. **Hybrid breakdown**—If the two different species produce offspring that are viable and fertile, these hybrids may mate to produce weak or sterile offspring.

Modes of Speciation

- There are two main types of speciation—**allopatric speciation**, in which a population forms a new species because it is geographically isolated from the parent population, and **sympatric speciation**, in which a small part of a population becomes a new population without being geographically separated from the parent population.
- Some **geologic events or processes** that can fragment a population include the emergence of a mountain range, the formation of a land bridge, or the evaporation of a large lake to produce several small lakes.
- Small, newly isolated populations undergo **allopatric speciation** more frequently because they are more likely to have their gene pools significantly altered. Allopatric speciation has occurred when an individual from the new population is unable to mate successfully with an individual from the parent population.
- **Adaptive radiation** occurs when many new species arise from a single common ancestor.
- One mechanism that can lead to the formation of a small new population within the parent population (**sympatric speciation**) in plants is the formation of **autopolyploid** plants through nondisjunction in meiosis. These plants have 4*n* chromosomes, instead of the normal 2*n* number, and they are unable to breed with members of the parent population—though they are still able to breed with other tetraploids.
- **Polyploid speciation** occurs in animals, but it is not common. Instead, in animals, sympatric speciation can result from part of the population switching to a new habitat, food source, or other resource.
- The **punctuated equilibrium model** states that species diverge in relatively quick spurts rather than slowly and gradually.

From Speciation to Macroevolution

- **Evo-devo** is the field where evolutionary biology and developmental biology meet.
- **Allometric growth** refers to the different growth rates of various parts of an organism's body during development.

- **Homeotic genes** determine the location and organization of body parts. Hox genes are one class of homeotic genes.

Phylogeny and Systematics: The Fossil Record and Geologic Time

- **Phylogeny** is the evolutionary history of a species or a group of related species, and **systematics** is the study of biological diversity in an evolutionary context.
- The **fossil record** is the sequence in which fossils appear in the layers of sedimentary rock that constitute the earth's surface. **Paleontologists** study the fossil record. Fossils are most often found in sedimentary rock, which are formed from layers of minerals settling out of water. Dead organisms settle along with the sediments and are compacted and sometimes preserved as fossils. Some fossils are not parts of organisms left behind, but impressions that the organisms made in sediment before they decayed and disappeared, as in a cast or an imprint. The fossil record is incomplete because it favors organisms that existed for a long time, were relatively widespread, and had shells or hard bony skeletons.
- **Relative dating** can be used to date fossils. The fossils in each layer of sedimentary rock represent certain time periods; the layers nearer the surface representing more recent time periods, and those further down representing more ancient times.
- **Index fossils** are those found in more than one location that can be used to correlate information about the strata in both locations.
- The **geologic time scale** shows a consistent sequence of periods and events in the history of earth. It is divided into four eras—**Precambrian, Paleozoic, Mesozoic,** and **Cenozoic.**

The Geologic Time Scale

Relative Time Span of Eras	Era	Period	Epoch	Age (Millions of Years Ago)	Some Important Events in the History of Life
Cenozoic	Cenozoic	Quaternary	Recent		Historical time
Mesozoic				0.01	
			Pleistocene		Ice ages; humans appear
Paleozoic				1.8	
		Tertiary	Pliocene		Apelike ancestors of humans appear
				5	
			Miocene		Continued radiation of mammals and angiosperms
				23	
			Oligocene		Origins of many primate groups, including apes
				35	
			Eocene		Angiosperm dominance increases; continued radiation of most modern mammalian orders
				57	
			Paleocene		Major radiation of mammals, birds, and pollinating insects
				65	
	Mesozoic	Cretaceous			Flowering plants (angiosperms) appear; many groups of organisms, including dinosaurs, become extinct at end of period (Cretaceous extinctions)
				144	
		Jurassic			Gymnosperms continue as dominant plants; dinosaurs abundant and diverse
				206	
		Triassic			Cone-bearing plants (gymnosperms) dominate landscape; radiation of dinosaurs
				245	
Pre-cambrian	Paleozoic	Permian			Extinction of many marine and terrestrial organisms (Permian mass extinction); radiation of reptiles; origins of mammal-like reptiles and most modern orders of insects
				290	
		Carboniferous			Extensive forests of vascular plants; first seed plants; origin of reptiles; amphibians dominant
				363	
		Devonian			Diversification of bony fishes; first amphibians and insects
				409	
		Silurian			Diversity of jawless fishes; first jawed fishes; diversification of early vascular plants
				439	
		Ordovician			Marine algae abundant; colonization of land by plants and arthropods
				510	
		Cambrian			Radiation of most modern animal phyla (Cambrian explosion)
				543	
	Precambrian			600	Diverse soft-bodied invertebrate animals; diverse algae
				2,200	Oldest fossils of eukaryotic cells
				2,700	Atmospheric oxygen begins to accumulate
				3,500	Oldest fossils of cells (prokaryotes)
				3,800	Earliest traces of life
				4,600	Approximate time of origin of Earth

- **Absolute dating** refers to fossils' age given in years, rather than given in reference to one another.
- **Radiometric dating** involves measuring the level of certain radioactive isotopes in fossils or rocks to determine their ages.
- The **half-life** of an isotope is the length of time it takes for 50% of the original sample of the isotope to decay. Paleontologists can measure the amount of an isotope in a rock or fossil, and then—knowing the isotope's half-life—use that measurement to determine the approximate age of the rock.
- The **continents** are not fixed; rather, they move slowly on tetonic plates of the earth's crust, which in turn float on the liquid mantle of the earth. About 250 million years ago, all of the major landmasses were brought together in a supercontinent called **Pangaea**; this event shaped biological evolution significantly. Pangaea broke up about 180 million years ago, causing extensive **geographic isolation** which eventually led to evolution.
- Extinction can occur because the habitat of a species is destroyed or because the environment has changed to make the species' existence impossible. There have been at least two major extinctions in the history of life on Earth—the **Permian extinction** and the **Cretaceous extinction**.

Systematics: Connecting Classification to Phylogeny

- **Systematics** is the study of biological diversity in the context of evolution, and it includes taxonomy, which is the naming and classification of species and groups of species.
- The **binomial** is used to describe species. It consists of the **genus** to which the species belongs, as well as the organisms' **species** within the genus. Organisms also have common names, like cat and dog.
- The hierarchical classification of organisms consists of the following levels (in order of increasing broadness): **species**, **genus**, **family**, **order**, **class**, **phyla**, **kingdom**, and **domain**. Each categorization at any level is called a **taxon**.

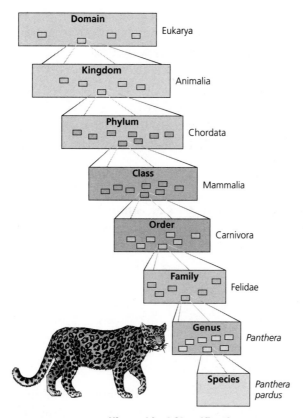

Hierarchical Classification

■ **Phylogenetic trees** show the hierarchical classification of taxonomic levels for organisms.

■ Relational diagrams based on phylogeny are called **cladograms**, and each branch of the diagram is called a **clade**. The sequence of branches on cladograms are determined on the basis of **homology**, which is likeness that is based on shared ancestry.

■ **Convergent evolution** has taken place when two organisms became alike because they adapted to similar environmental challenges—not because they have evolved from a common ancestor. The likenesses that result from convergent evolution are called **analogy** (not homology).

■ The greater the homology between two organisms, the more closely related they are, evolutionarily through common ancestry.

■ The comparison of genes and proteins (**molecular systematics**) of different organisms allows us to determine evolutionary relationships on a molecular level. The more alike the DNA sequences of two organisms are, the more closely they are related evolutionarily.

■ The rate of evolution of DNA sequences varies from one part of the genome to another, so comparing these different sequences helps us to investigate relationships between groups of organisms that diverged a long time ago.

■ The principle of **parsimony** dictates that theories should be kept as simple as possible while still being consistent with the evidence. This way of thinking is used to create **phylogenies**.

- Parsimony can be applied to the creation of **phylogenetic trees**, because in evolution heredity fidelity is more common than change (steps toward evolution). Phylogenetic trees are hypothetical.
- **Molecular clocks** are methods used to place the origin of taxonomic groups more precisely in time.

    For Additional Review: Analyze the underlying causes of evolution on a molecular scale, taking into account how changes in gene structure could cause alterations in the appearance of an organism and eventually lead to speciation.

Multiple-Choice Questions

1. The condition in which there are barriers to inbreeding between individuals of the same species separated by a portion of a mountain range is referred to as
   - (A) minute variations
   - (B) geographic isolation
   - (C) infertility
   - (D) reproductive isolation
   - (E) differential breeding capacity

2. Which of the following statements best expresses the concept of gradualism?
   - (A) Minute changes in the genome of individuals eventually leads to the evolution of a population.
   - (B) The three conditions of Hardy-Weinberg equilibrium will prevent populations from evolving quickly.
   - (C) Evolution occurs in rapid bursts of change alternating with long periods in which species remain relatively unchanged.
   - (D) Profound change over the course of geologic history is the result of an accumulation of slow, continuous processes.
   - (E) When two species compete for a single resource in the same environment, one of them will gradually become extinct.

3. A number of different phylogenies have been proposed by scientists over the decades. These are useful because they
   - (A) predict which species will evolve the most quickly in the future
   - (B) give us information about which species evolved most quickly in the past
   - (C) allow us to determine when two populations that are similar evolved into separate species
   - (D) show that all living creatures evolved from a common ancestor
   - (E) allow us to study evolutionary relationships and evaluate the relatedness of living organisms

4. All of the following statements are part of Darwin's theory of evolution EXCEPT
   - (A) the most prominent contribution to evolution is made by the process of genetic mutation
   - (B) natural selection is the force behind evolution
   - (C) natural selection occurs as a result of the differing reproductive success of individuals in a population
   - (D) the driving force of evolution is the adaptation of a population of organisms to their environment
   - (E) more individuals are born in a population than will survive to reproduce

5. In a certain group of rabbits, the presence of yellow fur is the result of a homozygous recessive condition in the biochemical pathway producing hair pigment. If the frequency of the allele for this condition is 0.10, which of the following is closest to the frequency of the dominant allele in this population? (Assume that the population is in Hardy-Weinberg equilibrium.)

(A) .01
(B) .20
(C) .40
(D) .90
(E) 1.0

Directions: The group of questions below consists of five lettered headings followed by a list of numbered phrases or sentences. For each numbered phrase or sentence select the one heading that is most closely related to it and fill in the corresponding oval on the answer sheet. Each heading may be used once, more than once, or not at all.

Questions 6–10

(A) Artificial selection
(B) Homology
(C) Gene pool
(D) The founder effect
(E) The bottleneck effect

6. Creates new species that possess certain desired traits

7. Can result in a new population with a limited gene pool

8. One result of evolution from a common ancestor

9. A result of drastic reduction in population size

10. Constitutes all of the alleles in a population

11. All of the following are examples of prezygotic barriers EXCEPT
(A) habitat isolation
(B) behavioral isolation
(C) temporal isolation
(D) mechanical isolation
(E) hybrid breakdown

12. Species that are found only in one particular geographic location are said to be
(A) behaviorally evolved from that location
(B) endemic to the location
(C) speciated
(D) undergoing behavioral isolation
(E) undergoing mechanical isolation

13. The allele that causes sickle-cell anemia is found with greater frequency in Africa, where malaria is more of a threat, than in the United States. This is due to which of the following genetic phenomena?
(A) Heterozygote advantage
(B) Heterozygote protection theory
(C) Balanced polymorphism
(D) Frequency-dependent selection
(E) Neutral variations

14. In a population of squirrels, the allele that causes bushy tail (B) is dominant, while the allele that causes bald tail is recessive (b). If we know the frequency with which the bushy tail allele occurs, we can calculate the frequency of the bald tail allele with which of the following equations?
(A) $b - B = 1$
(B) $2b = B$
(C) $b = 1 - B$
(D) $b^2 - 1 = B$
(E) $b + 1 = B$

15. The categories in which systematicists place species, in order of increasing specificity, are
    (A) species, genus, family, order, class, phyla, kingdom, domain
    (B) domain, kingdom, phyla, class, order, family, genus, species
    (C) class, domain, family, genus, kingdom, order phyla, species
    (D) family, genus, order, phyla, species, kingdom, domain, class
    (E) phyla, genus, order, species, class, domain, kingdom, order

16. A marsupial living in Australia has evolved to eat tree leaves, be diurnal, and raise its young until it is of reproductive age. A grazing mammal has also evolved to eat tree leaves, be diurnal, and raise its young until it is of reproductive age. This is an example of which of the following types of evolution?
    (A) Divergent evolution
    (B) Species-specific evolution
    (C) Convergent evolution
    (D) Neutral evolution
    (E) Sibling evolution

17. Which of the following can lead to sympatric speciation?
    (A) Migration of a small number of individuals
    (B) Natural disaster that cuts off contact between members of a population
    (C) Mutation
    (D) Autopolyploidy
    (E) Bottleneck effect

18. If the half-life of uranium-238 is 48 years, and if 25% of the original uranium sample present in a particular fossil was detected, approximately how old would the fossil be?
    (A) 50 years old
    (B) 100 years old
    (C) 200 years old
    (D) 250 years old
    (E) 500 years old

19. The supercontinent that existed before the continents broke apart and created the landforms we know today was called
    (A) Narnia
    (B) Octavia
    (C) Pangaea
    (D) Atlantis
    (E) Oz

20. Which of the following constitutes the smallest unit capable of evolution?
    (A) An individual
    (B) A group
    (C) A population
    (D) A clade
    (E) A community

Free-Response Question

1. *Microevolution is the change in the gene pool from one generation to the next.*

    ▌ Describe three different ways in which microevolution can take place.
    ▌ Describe the difference between microevolution and macroevolution.

ANSWERS AND EXPLANATIONS

Multiple-Choice Questions

1. (B) is correct. When two members of the same species are prevented from breeding by a geographic feature such as a mountain range or river, the fact that they live in different ponds, or any other physical obstruction that prevents them from meeting, the two individuals are said to be geographically isolated.

2. (D) is correct. The concept of gradualism was put forth by James Hutton in 1795. He stated that it was possible to understand the landforms that exist in the world today by looking at the mechanisms currently at work in the world today. This concept was meant to apply not to living creatures necessarily, but rather to Earth's geologic features. However, the concept influenced Darwin's thinking and may have contributed to the development of his theories.

3. (E) is correct. Phylogenies are evolutionary trees showing the history of a species or group of related species. Systematicists name and classify organisms, and create phylogenies based on the fossil record and accumulated physical and molecular data. Phylogenetic trees reflect the hierarchical classification of groups of species and of singular species, and they show how species and groups of species are related to one another.

4. (A) is correct. All of the answers are parts of Darwin's theory of evolution except *A*—genetic mutation is not the most important factor contributing to the process of evolution. In fact, when Darwin was developing his ideas, the existence of the gene was not yet known. Darwin's theory centered on his observation of changing populations rather than molecular evidence.

5. (D) is correct. If the frequency of the recessive allele is .10, then we know that the frequency of the other allele is 1 – .10, which is equal to .90. We know this because the Hardy-Weinberg equation states that, if a population contains just two alleles for a given trait, and we know the frequency of one of the alleles, we can calculate the frequency of the other using the equation $p + q = 1$. If we designate the frequency of the occurrence of the recessive allele as p, and use its value of .1, we can rearrange the equation to read $.10 + q = 1$. Then, $1 - .10 = .9$, which is equal to q, or the frequency of the other allele—in this case the dominant one.

6. (A) is correct. Artificial selection is the selective breeding of domesticated plants and animals in order to modify them to better suit the needs of humans. Humans have been practicing artificial selection for many generations, and many of the common foods we eat are a result of this process.

7. (D) is correct. The founder effect occurs when a few individuals from a population colonize a new, isolated habitat. The smaller the number of individuals who start this new population, the more limited will be the starting gene pool for the population—and the less the new gene pool will resemble that of the parent population.

8. (B) is correct. Homology is the result of descent from a common ancestor. It can be described as the underlying structural or molecular similarities (even in structures that are no longer used for the same function) that exist in organisms as a result of common ancestry.

9. (E) is correct. Bottleneck effects are often the result of a natural disaster such as a flood, drought, fire, or anything else that destroys most members of a population. The gene pool of the surviving members of the population may not resemble the gene pool of that of the parent population—some genes will be over-represented, and some will be underrepresented. Bottlenecking reduces the genetic variability in a population because of the loss of alleles.

10. (E) is correct. The gene pool is the collection of all of the alleles that exist in a population—a population is defined as a group of individuals living in a certain geographic location that are capable of interbreeding.

11. (E) is correct. Prezygotic barriers are those that prevent or hinder the mating of two different species, or they prevent fertilization of ova if two species are able to mate. All of the answers are examples of prezygotic barriers except the last one. Hybrid breakdown is an example of a postzygotic barrier (Postzygotic barriers are those that prevent hybrid zygotes from developing into viable adults), in which the second generation of offspring from two different species are either weak or sterile.

12. (B) is correct. Species that are found only in one geographic location are said to be endemic to that location. Some examples of endemic species are kangaroos (endemic to Australia) and blue-footed boobies (endemic to the Galapagos Islands).

13. (A) is correct. Individuals heterozygous for the allele that causes sickle-cell anemia are favored in Africa—this phenomenon is known as heterozygote advantage. The reason they are favored is because homozygotes for sickle-cell anemia have the disease, which can lead to death, while those homozygous for the regular, non-sickle-cell-causing allele have less resistance to malaria. Thus, some of those individuals die from malaria.

14. (C) is correct. We can calculate the frequency of the recessive allele for bald tail in squirrels using the equation $b = 1 - B$, where B is the frequency of the allele that causes bushy tail in squirrels. This is just a rearrangement of the Hardy-Weinberg equation, which in this case would read $B + b = 1$. If there are only two alleles for a given trait in a population, their gene frequencies will always add to 1.

15. (B) is correct. The categories in which systematicists place organisms—in order of increasing specificity—is domain, kingdom, phyla, class, order, family, genus, and species. This means that the broadest category is the domain, and according to the existing school of thought, there are three domains of living creatures—archaea, eubacteria, and eukarya. The second largest and most inclusive group under domain is kingdom, and so on.

16. (C) is correct. In convergent evolution, species from different evolutionary branches come to appear alike as a result of undergoing evolution in very similar ecological roles and environments. Similarity between species that have undergone convergent evolution is known as analogy, and structures they share in common are analogous (not homologous) structures. Homologous structures are those that have evolved in two different species as a result of those species' having a common ancestor.

17. (D) is correct. One mechanism that can lead to sympatric speciation, which is the formation of a small new population within the parent population in plants, is the formation of autopolyploids through nondisjunction in meiosis. These plants have $4n$ chromosomes, instead of the normal $2n$ number, and they are unable to breed with members of the parent population—though they are still able to breed with other tetraploids.

18. (B) is correct. The fossil that was found would be 98 years old. If the sample started with 100% uranium-238 (which is what we can assume from the question), then after 48 years, 50% of the uranium would be left, and after another 48 years, half of that—or about 25% of the original uranium—would be left. The half-life is a measure of the amount of time it takes for half of a radioactive sample to decay.

19. (B) is correct. The massive continent that existed before the smaller continents that we know today broke off and drifted into their current positions was known as Pangaea. The continents are capable of floating around as they do because they are located on gigantic plates, which in turn ride along on the molten mantle of the earth.

20. (C) is correct. The smallest unit capable of evolution is the population. Individuals cannot undergo evolution, because they only exist for one generation, and evolution is the changing and refinement of a group's gene pool to best fit the group's environment.

Free-Response Question

1. Microevolution is the change in the gene pool of a population that occurs from one generation to the next, and three factors that contribute to microevolution are genetic drift (including both the bottleneck effect and the founder effect), natural selection, and gene flow.

   The term *genetic drift* refers to any change in a population's allele frequencies due to chance. Two examples of what this "chance" can consist of are the bottleneck effect and the founder effect. The bottleneck effect occurs after a natural disaster such as a violent storm, fire, or other catastrophe causes a drastic reduction in the size of the population. Such an event leaves only a few individuals to continue to produce offspring, so bottlenecking usually reduces the genetic variability in a population, since some alleles are lost from the gene pool. Similarly, the founder effect occurs when a few members of a population colonize an isolated location that isn't accessed by members of the parent population. The smaller the number of founders, the more limited the variability of the genes in the new population.

   Natural selection is another route by which microevolution can take place. This term refers to the differing reproductive success of all of the individuals in a population. Those who are best suited to their environment will survive to pass on their alleles to the next generation.

   Gene flow refers to genetic exchange due to the migration of individuals or gametes between populations. This can take place in the course of one generation to the next, so this is another valid contributor to microevolution.

Macroevolution is the process of the creation of new taxonomic groups (meaning, new species, families, or kingdoms) through evolution. It differs from microevolution in that microevolution refers only to changes that occur in populations from generation to generation—no new species or other taxonomic groups need arise in the course of microevolution. Microevolution is small-scale, whereas macroevolution concerns the "bigger picture."

This is a good free-response answer because it shows knowledge of the following key terms that you will be expected to know for the exam:

microevolution	*founder effect*
gene pool	*natural selection*
population	*gene flow*
genetic drift	*macroevolution*
bottleneck effect	*taxonomic groups*

It also shows that the student understands the following processes: microevolution, genetic drift, the bottleneck effect and the founder effect, natural selection, gene flow, and macroevolution.

CHAPTER 5

The Evolutionary History of Biological Diversity

Early Earth and the Origin of Life: Introduction to the History of Life

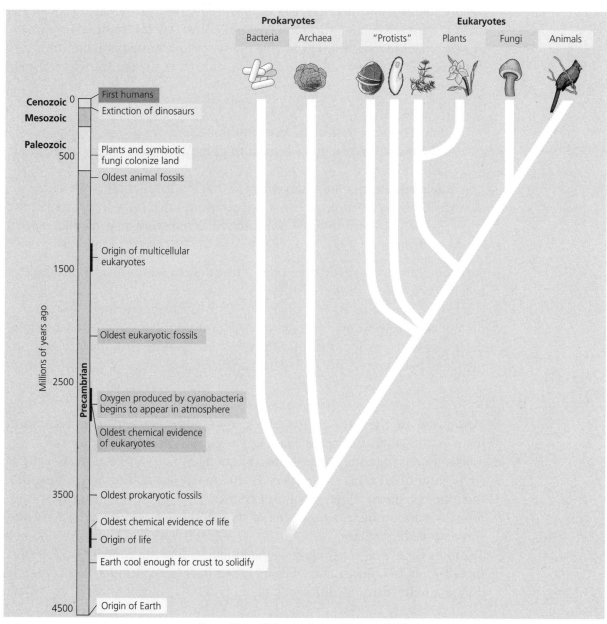

Some Major Episodes in the History of Life

- Life on Earth started about **3.5 to 4 billion years ago**, but for the first three-quarters of Earth's history, all of its living organisms were microscopic and primarily unicellular.
- The earliest living organisms were **prokaryotes**.
- About 2.7 billion years ago, **oxygen** began to accumulate in the atmosphere of Earth.
- **Eukaryotes** appeared about 2.1 billion years ago, and **multicellular eukaryotes** evolved about 1.2 billion years ago.
- About 500 million years ago, **plants, fungi** and **animals** began to appear on Earth.

The Origin of Life

- Most scientists agree that life on Earth began after **nonliving materials** became organized into **molecular aggregates** that eventually became capable of reproducing themselves and metabolizing molecules. In the past, many people thought that life arose from nonliving matter by **spontaneous generation**.
- **Biogenesis** is the theory that life can arise only from the reproduction of pre-existing life.
- Current theories about how life arose consists of four parts:

  1. Small organic molecules were synthesized.
  2. These small molecules joined into polymers, like proteins and nucleic acids.
  3. Self-replicating molecules emerged that made inheritance possible.
  4. All of these molecules were packaged into membrane-containing droplets, whose internal chemistry differed from that of the external environment.

- Early conditions of Earth have been simulated in laboratories, and organic polymers have been produced.
- It is hypothesized that RNA (not DNA) was the **first genetic material**.
- Before cells, **protobionts** may have existed. Protobionts are aggregates of molecules that were produced abiotically; they have a consistent internal environment and have some other properties associated with life.

The Major Lineages of Life

- Introduced in the early 1970s, the **five-kingdom system** of classification includes the Monera, Protista, Plantae, Fungi, and Animalia kingdoms.
- Relatively recently, the **three-domain system**, which consists of Eubacteria, Archaea, and Eukarya, was proposed. This system arose from the finding that there are two distinct lineages of prokaryotes. This system also makes the kingdom Monera obsolete, since some of its members would be Eubacteria and some would be Archaea.
- **Taxonomy** is a work in progress—the kingdom Protista is now being studied more closely, too, since many of its members have been found to be sufficiently different to be placed in different kingdoms.

Prokaryotes and the Origins of Metabolic Diversity: The World of Prokaryotes

Life is divided into three domains: **Archae**, **Eubacteria**, and **Eukarya**. Both domain Eubacteria and domain Archae are made up of prokaryotes.

The Structure, Function, and Reproduction of Prokaryotes

- The most common shapes of prokaryotes are spheres, rods, and helices; most are about 1–5 µm in size.
- Outside their cell membranes, most prokaryotes possess a cell wall that contains **peptidoglycans**. **Gram-positive** bacteria have simpler walls with more peptidoglycan, whereas **gram-negative** cells have walls that are structurally more complex.
- Prokaryotes use appendages called **pili** to adhere to each other or to surrounding surfaces. About half of the prokaryotes are **motile**, because they possess a whip-like **flagella**. Prokaryotes do not have true nuclei or internal compartmentalization. Relative to eukaryotes, prokaryotes have simple, small genomes. The DNA is concentrated in a nucleoid region and has little associated protein.
- In addition to their one major chromosome, prokaryotic cells may also possess smaller, circular, independent pieces of DNA called **plasmids**.
- Prokaryotes reproduce through an asexual process called **binary fission**, and they continually synthesize DNA.
- Three mechanisms by which bacteria can transfer genetic material between each other are:

  - **transformation**, in which a prokaryote takes up genes from its environment
  - **conjugation**, in which genes are directly transferred from one prokaryote to another
  - **transduction**, in which viruses transfer genes between prokaryotes

- The major source of genetic variation in prokaryotes is **mutation**.

Nutritional and Metabolic Diversity

- Prokaryotes can be placed in four groups according to how they take in carbon and how they provide themselves with energy.

  - **Photoautotrophs** are photosynthetic, and they use the power of sunlight to turn carbon dioxide into organic compounds.
  - **Chemoautotrophs** also use carbon dioxide as their source of carbon, but they get energy from oxidizing inorganic substances.
  - **Photoheterotrophs** use light to make ATP but must obtain their carbon from an outside source already fixed in organic compounds.
  - **Chemoautotrophs** get both carbon and energy from organic compounds.

- Most prokaryotes are chemoheterotrophs, and there are two types of chemoheterotrophs: **saprobes** (which are decomposers that absorb nutrients from dead organic matter) and **parasites** (which absorb nutrients from the body fluids of living hosts).

- Some prokaryotes can use atmospheric nitrogen as a direct source of nitrogen in a process called **nitrogen fixation**. They convert N_2 to NH_4^+.
- **Obligate aerobes** cannot grow without oxygen. They use oxygen for cell respiration. **Obligate anaerobes** use fermentation, and are poisoned by oxygen. **Facultative anaerobes** use oxygen if it is available; when no oxygen is available, they undergo fermentation.

A Survey of Prokaryotic Diversity

- Prokaryotes are very diverse and are able to live in extreme environments. **Extremophiles** live in extreme environments such as geysers.
- There are three types of extremophiles. **Methanogens** use carbon dioxide to oxidize H_2 and to produce methane as a byproduct; **extreme halophiles** live in saline environments (highly concentrated with salt); and **extreme thermophiles** live in very hot environments.
- Most known prokaryotes are bacteria.

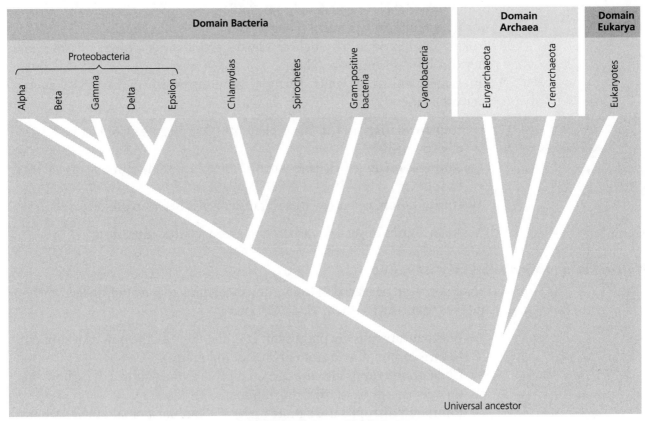

Some Major Groups of Prokaryotes

The Ecological Impact of Prokaryotes

▌ Prokaryotes are responsible for the **decomposition of living matter**, and this is the first step in the cycling of essential elements through the environment.

▌ Many prokaryotes are **symbiotic**, meaning that they form relationships with other species. In symbiotic relationships where one organism is significantly larger than the other, the larger one is called the **host**. There are three types of symbiosis.

   1. **mutualism**—in which both symbiotic organisms benefit
   2. **commensalism**—in which one organism benefits while the other is neither helped nor harmed
   3. **parasitism**—in which one organism benefits while the other is harmed or eventually can die

The Origins of Eukaryotic Diversity: Introduction to the Protists

▌ **Protists** are the simplest but most diverse of all eukaryotes. They vary in structure and function more than any other group of eukaryotes, although most are unicellular. Most protists use aerobic metabolism, and have mitochondria.

▌ The animal-like protists are called **protozoa**, whereas the plant-like protists are called **algae**.

▌ Most protists have cilia or flagella. Most protists reproduce through mitosis. Most protists live in aquatic environments.

The Origin and Early Diversification of Eukaryotes

▌ Eukaryotes have internal membranes and compartmentalization, and these features make them much more complex than prokaryotes.

▌ According to current theory, mitochondria and chloroplasts are descended from endosymbiotic bacteria that took up residence in eukaryotic cells that were evolving.

A Sample of Protistan Diversity

A Sample of Protistan Diversity		
Major Clade	**A Key Characteristic**	**Example**
❶ Diplomonadida-Parabasala	Secondary loss of mitochondria	
A. Diplomonadida (Diplomonads)	Two separate nuclei	*Giardia*
B. Parabasala (Trichomonads and other parabasalids)	Undulating membrane	*Trichomonas*
❷ Euglenozoa	Photosynthetic, heterotrophic, and mixotrophic flagellates	
A. Euglenophyta (euglenoids)	Paramylon as storage polysaccharide	*Euglena*
B. Kinetoplastida (kinetoplastids)	Kinetoplast, a unique organelle	*Trypanosoma*
❸ Alveolata	Subsurface alveoli (membrane-bound cavities)	
A. Dinoflagellata (dinoflagellates)	Armor of cellulose plates	*Pfiesteria*
B. Apicomplexa (apicomplexans)	Apical complex functioning in penetration of host cells	*Plasmodium*
C. Ciliophora (ciliates)	Cilia functioning in movement and feeding	*Paramecium*
❹ Stramenopila	"Hairy" flagella	
A. Oomycota (oomycetes)	Hyphae that absorb nutrients	Water molds, rusts, downy mildews
B. Bacillariophyta (diatoms)	Glassy, two-part walls	*Pinnularia*
C. Chrysophyta (golden algae)	Biflagellate cells; xanthophyll pigments	*Dinobryon*
D. Phaeophyta (brown algae)	Brown color from accessory pigments	*Laminaria*
❺ Rhodophyta (red algae)	No flagellated stages; phycoerythrin pigment	*Porphyra* ("Nori")
❻ Viridiplantae (includes the green algal group Chlorophyta)	Plant-type chloroplasts	*Chlamydomonas*
❼ Mycetozoa	Decomposers having complex life cycles with amoeboid stages	
A. Myxogastrida (plasmodial slime molds)	Netlike plasmodium as feeding stage	*Physarum*
B. Dictyostelida (cellular slime molds)	Amoeboid feeding cells that aggregate to form reproductive colonies	*Dictyostelium*
Pseudopod-equipped protists of uncertain phylogeny	Pseudopodia that function in movement and feeding	
A. Rhizopoda (rhizopods)	Lobe-like pseudopodia	*Amoeba*
B. Actinopoda (actinopods)	Ray-like pseudopodia (axopodia)	Heliozoans and radiolarians
C. Foraminifera (forams)	Porous shells	*Globigerina*

Plant Diversity I: How Plants Colonized Land: An Overview of Land Plant Evolution

▌ Plants are multicellular, eukaryotic, photosynthetic autotrophs.

▌ There are four main groups of land plants: **bryophytes**, **pteridophytes**, **gymnosperms**, and **angiosperms**.

▌ **Bryophytes** live on land and have several adaptations for land-living that distinguishes them from algae. They lack vascular tissues, which pteridophytes, gymnosperms, and angiosperms have.

▌ **Pteridophytes** lack seeds, which both gymnosperms and angiosperms have.

▌ **Seeds** are plant embryos packaged with a food supply in a protective coat.

- **Gymnosperms** have seeds that are not enclosed in a protective coating.
- **Angiosperms** are flowering plants.
- All land plants have a life cycle that consists of two stages, called **alternation of generations**. The two multicellular stages are the **gametophyte** stage (in which the plant cells are haploid) and the **sporophyte** stage (in which the plant cells are diploid).
- In the gametophyte stage, the gametes are produced, and in fertilization egg and sperm fuse to form a diploid zygote, the sporophyte, which divides mitotically.
- A **spore** is a cell produced by the plant that can develop into a new plant without fusing with another cell.
- Plants that live on land have many adaptations that enable them to do so.

    1. Adaptations for the conservation of water include the cuticle on the epidermis. The **cuticle** is a waxy layer made up of polymers that seals water into the cell. Another adaptation is the **stomata**. These openings on the undersurface of the leaf open and close to allow the passage of CO_2 and H_2O.
    2. Some adaptations for the transport of water through the body of the plant are **xylem** and **phloem**. These tissues are made up of dead cells through which water is conducted from stem to roots and vice versa.
    3. The adaptation for the transport of photosynthesized food (glucose) are the **phloem** tubules. They will conduct glucose from the leaf down the stem to the roots.

- Land plants evolved from algae over 500 million years ago. Evidence for this event includes the similar (homologous) chloroplasts, cellulose walls, peroxisomes, and sperm of these organisms.

Bryophytes

- There are three phyla in **bryophyta—mosses, liverworts,** and **hornworts.**
- The dominant generation in bryophytes is the **gametophyte**, and bryophytes disperse spores in great amounts.
- **Mosses** are one prominent example of bryophytes.

Pteridophytes: Seedless Vascular Plants

- The seedless vascular plant life cycle is dominated by the **sporophyte** stage.
- **Ferns** and **horsetails** are prominent examples of pteridophytes.
- The seedless vascular plants formed great forests in the Carboniferous period, eventually forming deposits of coal that we use for fuel today.

Plant Diversity II: The Evolution of Seed Plants: Overview of Seed Plant Evolution

- **Seed plants** are defined as vascular plants that produce seeds.
- The three most crucial adaptations of plants that led to the evolution of seed plants are:

    1. reduction of the gametophyte stage
    2. evolution of the **seed**
    3. evolution of **pollen**

- The microspores of seed plants develop into pollen grains, which are dispersed by the wind or animals, so that the process of **pollination** can take place.
- **Gymnosperms** and **angiosperms** are the two types of seed plants.

Gymnosperms

- The prominent gymnosperms are the **conifers**, which are cone-bearing plants like pine trees.
- Pine trees bear both pollen cones and ovulate cones.

Angiosperms

- **Angiosperms** are flowering plants, and all of them are placed into the phyla **Anthophyta**.
- The **monocots** are generally described as having veins that run parallel on the surface of a leaf, whereas **dicots** have netlike vein patterns on its leaves.
- The major reproductive adaptation of the angiosperm is the **flower**.
- The flower is specialized for reproduction.
- **Fruits** are mature ovaries of the plant. They help disperse the seeds of angiosperms.
- The life cycle of the angiosperm is a refined version of the alternation of generations that all plants undergo.

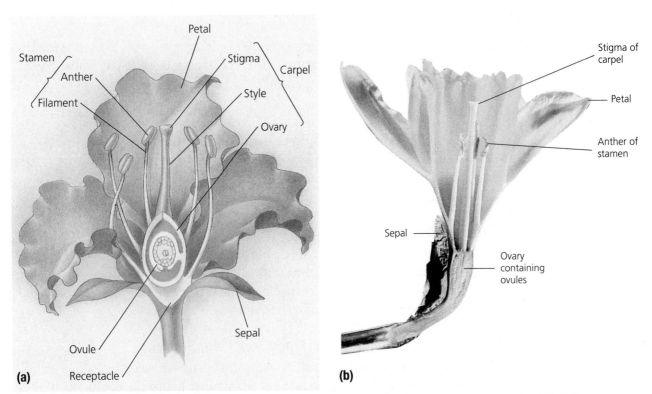

The Structure of a Flower. (a) The parts of an idealized flower. (b) The photo shows a cutaway of a daffodil flower.

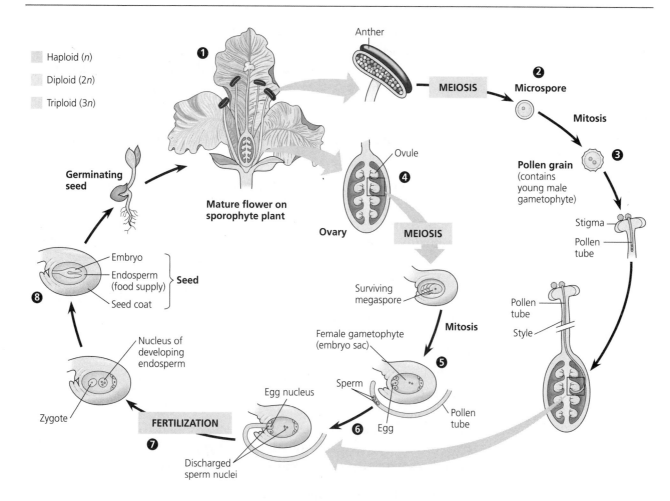

The Life Cycle of an Angiosperm

Legend:
- Haploid (*n*)
- Diploid (*2n*)
- Triploid (*3n*)

Diagram labels:
- ❶ Mature flower on sporophyte plant
- Anther
- MEIOSIS
- ❷ Microspore
- Mitosis
- ❸ Pollen grain (contains young male gametophyte)
- Stigma
- Pollen tube
- Pollen tube
- Style
- Ovule
- ❹ Ovary
- MEIOSIS
- Surviving megaspore
- Mitosis
- Female gametophyte (embryo sac)
- ❺
- Sperm
- ❻ Egg
- Pollen tube
- Egg nucleus
- FERTILIZATION
- ❼
- Discharged sperm nuclei
- Nucleus of developing endosperm
- Zygote
- ❽ Embryo
- Endosperm (food supply)
- Seed coat
- **Seed**
- **Germinating seed**

▌ The angiosperm life cycle consists of the following steps (see the diagram above): (1) The anthers of the flower produce (2) microspores. The microspores form (3) male gametophytes—also known as pollen. (4) Meanwhile, the ovules produce megaspores that form (5) female gametophytes (embryo sacs). (6) Pollination brings the gametophytes together in the ovary. (7) Double fertilization takes place, and (8) zygotes develop into sporophyte embryos that are packaged along with food into seeds.

▌ Animals and plants have **coevolved**, meaning that they have had a mutual evolutionary influence on each other.

Fungi: Introduction to the Fungi

▌ Most **fungi** are multicellular, and all are eukaryotes. They are different from other eukaryotes in their nutrition, structure, growth, and reproduction.

▌ Fungi are **heterotrophs**, and they obtain nutrients by **absorption**, in which they take up small molecules from their surroundings. Fungi secrete hydrolytic enzymes and digest food outside their bodies.

▌ The bodies of fungi are composed of filaments called **hyphae** that are entwined together to form a **mycelium** (the body).

- Most fungi are multicellular, and the hyphae are made up of cells with walls between them called **septa**. The cell walls are made up of chitin.
- Fungi produce spores either sexually or asexually, and they disperse these spores in order to reproduce. The spores of most fungal species are haploid.

Diversity of Fungi

- There are **four phyla of fungi**: Chytridiomycota, Zygomycota, Ascomycota, and Basidomycota.

    1. **Chytridomycota** (chytrids) are aquatic saprobes or parasites; they are thought to be the most primitive fungi.
    2. **Zygomycota** (zygote fungi) are terrestrial, and some form mycorrhizae, mutualistic associations with plant roots. A common zygomycote is bread mold, Rhizopus.
    3. **Ascomycota** (sac fungi) live in a variety of habitats.
    4. **Basidomycota** (club fungus) include mushrooms and are important decomposers of plant material.

- Three other important fungi are **molds, yeasts,** and **lichens**.

    - **Molds** are rapidly growing fungi that reproduce asexually.
    - **Yeasts** are unicellular fungi that live in moist habitats. They produce asexually by budding.
    - **Lichens** are symbiotic associations of photosynthetic microorganisms (algae) embedded in a network of fungal hyphae. They are very hardy organisms that are pioneers on rock and soil surfaces.

- Fungi are important decomposers that release inorganic nutrients, enabling them to cycle through the environment.

Introduction to Animal Evolution: What Is an Animal?

- Animals are **multicellular eukaryotes** that must ingest preformed organic molecules into their bodies. (They are heterotrophs.)

Two Views of Animal Diversity

There are about **35 animal phyla**. The following are important terms relating to the classification of animals.

- **Radial symmetry** occurs in jellyfishes and other organisms. Any cut through the central axis of the organism would produce mirror images.
- **Bilateral symmetry** is seen in lobsters, humans, and many other organisms. These animals have a right side and a left side, and a single cut would divide the animal into two mirror image halves.
- **Cephalization** is the concentration of sensory equipment at one end (usually the anterior or head end) of the organism.
- **Acoelomates**, such as flatworms, have no cavities between their digestive tracts and the outer wall of their bodies.
- **Coelomates** possess a **body cavity** filled with fluid, and this space separates an animal's digestive tract from the outer body wall.

The Origins of Animal Diversity

❚ The evolutionary episode that led to most of the animal phyla occurred during the **late Precambrian and early Cambrian eras**.

❚ During the Cambrian explosion, the first animals that possessed **hard, mineralized skeletons** appeared in the fossil record.

Invertebrates

Invertebrates are animals that lack backbones. An outline of animal diversity and classification follows.

1. **Subkingdom Parazoa: Phylum Porifera**

❚ **Parazoa** (sponges) are the oldest animals. They are sessile but very sedate and have no nerves or muscles. The body of a sponge looks like a sac with holes in it. Water is drawn through the pores into the **spongocoel** and flows out through the **osculum**.

❚ Most sponges are **filter-feeders** that collect particles from water that passes through them.

❚ Most sponges are **hermaphrodites**. That is, they function as both male and female and produce both sperm and eggs.

❚ Sponges are capable of **regeneration** of lost parts.

2. **Subkingdom Eumetazoa**

 ▪ **A. Radially symmetrical animals**
 All animals except the Parazoa belong to the clade Eumetazoa; they are animals with true tissues. There are two phyla of radiata—**cnidaria** and **ctenophora**.

 1. **Phylum Cnidaria** exist in polyp and medusa form, and they have radial symmetry, a central digestive compartment known as a gastrovascular cavity, and cnidocytes (cells that function in defense and the capture of prey). Examples of cnidarians are hydras, jellyfish, and sea anemones.
 2. **Phylum Ctenophora**, also known as comb jellies, look like medusal cnidarians. Most are spherical and possess rows of plates formed from fused cilia, which they use for locomotion. They also possess colloblasts, which are used in the capture of prey.

 ▪ **B. Bilaterally symmetrical animals**
 1. **Acoilomates (animals without a body cavity)**
 • **Phylum Platyhelminthes (flatworms)** live in water or damp terrestrial habitats. They exist in parasitic and free-living form, are flat with dorsal and ventral surfaces, lack organs that are specialized for gas exchange/circulation, and reproduce asexually through regeneration. Platyhelminthes are divided into four classes—Turbellaria, Monogenea, Trematoda, and Cestoidea. Some examples are flukes and tapeworms.

 • **Phylum Nemertea**, also called ribbon worms or proboscis worms, have excretory, sensory, and nervous systems similar to flatworms.

They have a complete digestive tract with a closed circulatory system but no heart.

2. Pseudocoelomates

- **Phylum Nematoda**, or roundworms, are found in most aquatic habitats. Their bodies are not segmented but are cylindrical. Roundworms have an exoskeleton called a cuticle, a complete digestive tract (but no circulatory system), and a pseudocoelom. They reproduce sexually, and sexes are separate in most species. Some examples are pinworms and hookworms.

- **Phylum Rotifera** inhabit fresh water. They have specialized organ systems, including a complete digestive tract and jaws that grind food and cilia that draw water into the mouth. Some species participate in parthenogenesis.

3. Coelomates

- **Phylum Lophophorata** includes Bryozoa (which look like mosses), Phoronida (marine worms), and Brachiopoda (lamp shells that look like clams). They contain a lophophore—a circular fold of the body wall with ciliated tentacles surrounding the mouth.

- **Phylum Mollusca** are soft-bodied animals protected with a hard shell (except for slugs, squids, and octopuses). They possess a muscular foot for movement, a visceral mass made up of the organs, and a mantle, which drapes over the visceral mass and secretes a shell. Most have separate sexes (though many snails are hermaphrodites), and most are marine. Some examples are Polyplacophora (chitons), Gastropoda (snails and slugs), Bivalva (clams and oysters), and Cephalopoda (squids and octopuses).

- **Phylum Annelida** are worms (such as the earthworm) that are segmented both externally and internally. They live in the sea, in fresh water, or in damp habitats. They have a coelom, a closed digestive system with specialized regions (the crop, gizzard, esophagus, and intestine), a brain-like central ganglia, and a hydrostatic skeleton that enables them to move. They can be hermaphrodites. Some examples are Oligochaeta (earthworms and their relatives), Polychaeta (polychaetes), and Hirudinea (leeches).

- **Phylum Arthropoda** are segmented animals with a hard exoskeleton and jointed appendages. In order to grow, arthropods must occasionally shed their exoskeleton and secrete a new one—this process is known as molting. They have well-developed sensory organs, an open circulatory system, organs specialized for gas exchange, and a homocoel (body sinuses functioning in transport of fluid through the body). Some examples are Crustacea (lobsters and shrimp), Chelicerates (spiders), insects , Trilobites (extinct), scorpions, millipedes, and centipedes.

4. Deuterostomia

The clade Deuterostomia contains a diverse array of organisms, from sea stars to chordates. All have radial cleavage and share common developmental processes. There are two main phyla.

- **Phylum Echinodermata**, or echinoderms, are slow-moving and radiate from the center. They have a thin skin covering an exoskeleton, as well as a water vascular system (a network of internal canals that branch into tube feet used for moving, feeding, and gas exchange). They reproduce sexually and can be divided into six classes. Some examples are sea stars, brittle stars, sea urchins, sand dollars, and sea cucumbers.

- **Phylum Chordata** includes two subphyla composed of invertebrates as well as *all* vertebrates. They are similar to one another developmentally.

Animal Phyla			
Category	**Phyla**		**Description of Phyla**
Kingdom Animalia Parazoa	Porifera (sponges)		Choanocytes (collar cells—unique flagellated cells that ingest bacteria and tiny food particles); cells tend to be totipotent (retain zygote's potential to form the whole animal)
Eumetazoa Radiata	Cnidaria (hydras, jellies, sea anemones, corals)		Unique stinging structures (cnidae), each housed in a specialized cell (cnidocyte); gastrovascular cavity (incomplete digestive tract with a mouth but no anus)
	Ctenophora (comb jellies)		Colloblasts (adhesive structures) for prey capture; eight rows of comblike ciliary plates; gastrovascular cavity
Bilateria Protostomia: Lophotrochozoa	Platyhelminthes (flatworms)		Dorsoventrally flattened, unsegmented acoelomates; gastrovascular cavity or no disgestive tract
	Rotifera (rotifers)		Pseudocoelomates with complete digestive tracts; jaws in pharynx structures (trophi); head with a cilated crown (corona); no circulatory system
	Lophophorates: Bryozoa, Brachiopoda, Phoronida		Coelomates with lophophore (feeding structure bearing cilated tentacles)
	Nemertea (proboscis worms)		Unique anterior proboscis surrounded by fluid-filled cavity (rhynchocoel); complete digestive tract (mouth and anus); circulatory system with closed vessels
	Mollusca (clams, snails, squids)		Coelomates with three main body parts (muscular foot, visceral mass, mantle); coelom reduced; main body cavity is a hemocoel
	Annelida (segmented worms)		Coelomates with body wall and internal organs (except digestive tract) segmented
Protostomia: Ecdysoza	Nematoda (roundworms)		Cylindrical, unsegmented pseudocoelomates with tapered ends; no circulatory system
	Arthropoda (crustaceans, insects, spiders)		Coelomates with segmented body, jointed appendages, exoskeleton from ectoderm
Deuterostomia	Echinodermata (sea stars, sea urchins)		Coelomates with secondary radial anatomy (larvae bilateral; adults radial); unique water vascular system; endoskeleton
	Chordata (lancelets, tunicates, vertebrates)		Coelomates with notochord; dorsal hollow nerve cord; pharyngeal slits; muscular postanal tail

Vertebrate Evolution and Diversity: Invertebrate Chordates and the Origin of Vertebrates

▌ **Vertebrates** are animals that have a backbone.

▌ There are **four anatomical features** that characterize chordates, many of which appear only during embryonic development in some chordates.

1. A **notochord**—a long flexible rod that appears during embryonic development between the digestive tube and the dorsal nerve cord
2. A **dorsal, hollow nerve cord**—formed from a plate of ectoderm that rolls into a hollow tube
3. **Pharyngeal slits**—allow water to enter and exit the mouth without going through the digestive tract
4. A muscular **tail** posterior to the anus

▌ There are two subphyla of invertebrate chordates—**Urochordata** (tunicates) and **Cephalochordata** (lancelets). They are simpler versions of vertebrates.

Introduction to the Vertebrates

Four things that differentiate the vertebrates are a **neural crest, significant cephalization, a vertebral column**, and a **closed circulatory system**

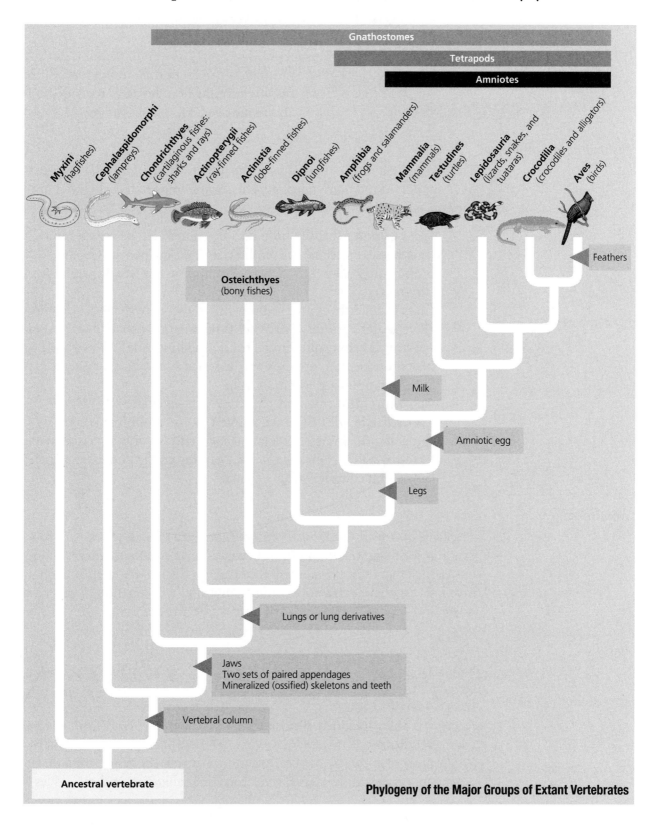

Phylogeny of the Major Groups of Extant Vertebrates

Jawless Vertebrates

Jawless vertebrates (agnathans) consist of two extant classes.

1. **Class Myxini** (hagfishes) are marine-dwellers and bottom-dwelling scavengers. They possess no vertebrae and have a skeleton made of cartilage.
2. **Class Cephalaspidomorphi** (lampreys) live in marine and fresh water. A cartilaginous pipe surrounds the notochord; lampreys lack skeleton-supported jaws. Their rasping mouths bore holes in the side of a fish. They live on the blood and tissue of a host.

Fishes and Amphibians

The jaws of vertebrates evolved from the modification of skeletal parts that had once supported the pharyngeal gill slits. Some important fishes and amphibians are listed below.

1. **Class Chondrichthyes** have flexible endoskeletons composed of cartilage; possess streamlined bodies; are denser than water and will sink if they stop swimming. Some examples are sharks and rays (like stingrays).
2. **Class Osteichthyes** are the bony fishes; these are the most numerous of all vertebrate groups. There are three main classes of bony fishes (the ray-finned fishes, lobe-finned fishes, and lungfishes). They have an ossified endoskeleton, are covered in scales, and possess a swim bladder. Some examples are common fish.
3. **Class Amphibia** maintain close ties with water and rely on their skin for gas exchange with the environment. Some amphibians have legs, and some do not. Some have a larval stage (frogs) with a dual (aquatic and terrestrial) life. Their eggs lack a shell, and fertilization is external. They can exhibit complex social behavior.

Amniotes

- The clade of **Amniotes** consists of mammals, birds, and reptiles.
- The **amniotic egg** was an important evolutionary development. Amniotic eggs have a shell that retains water and thus can be laid in a dry environment.
- Amniotic eggs have **extraembryonic membranes** that function in gas exchange, waste storage, and the transport of nutrients to the embryo.

Reptiles

- Reptiles have **scales**, which are an adaptation for terrestrial living. They obtain oxygen through their **lungs**, not their skin. They lay eggs on **land**; and undergo **internal fertilization**.
- Ancient reptiles included **dinosaurs** (which lived on land) and **pterosaurs** (flying reptiles).
- **Modern reptiles** consist of the Testudines (turtles), Sphenodontia (tuataras), Squamata (lizards and snakes), and Crocodilia (alligators and crocodiles).

Birds

- Birds lay **amniotic eggs and** have **scales** on their legs—both of which are vaguely reptilian.
- Most birds' bodies are constructed for **flight**, with light, hollow bones, relatively few organs, wings, and feathers. Flightless birds are called **ratites**, and birds that fly are called **carinates**.
- Birds are **endotherms** and maintain a warm, consistent body temperature. **Feathers** and, in some cases, a layer of fat insulate birds and help them maintain internal temperature.
- Birds have a **four-chambered heart** and a high rate of **metabolism**, and they have larger **brains** (compared proportionally) than reptiles do.

Mammals

- Most or all mammals share certain characteristics. For instance, all possess **mammary glands** that produce milk; all possess **hair**; all are **endothermic** and have an active metabolism; most are **born** rather than hatched; all use **internal fertilization**; all have **proportionally larger brains** than other vertebrates; and all have **teeth** of differentiating size.
- Mammals can be place into three groups.

    1. **Monotremes** are egg-laying mammals that have hair and produce milk. Some examples are platypuses and spiny anteaters.
    2. **Marsupials** are born early in development and complete embryonic development in a marsupian (pouch) while nursing. Examples include wombats, Tasmanian devils, and kangaroos.
    3. **Placental mammals** have a longer period of pregnancy; they complete their development in the uterus. Examples include deer mice, moles, flying squirrels, and wolverines.

Primates and the Evolution of Homo sapiens

- Some characteristics common to all **primates** include hands and feet that grasp; large brains and short jaws; forward-looking eyes; flat nails; well-developed parental care, and complex social behavior.
- The two subgroups of primates are **Prosimii** (lemurs, lorises, pottos and tarsiers) and **Anthropoidea** (monkeys, apes, and humans).
- Some features of **human evolution** are increased brain volume; shortening of the jaw; bipedal posture; reduced size-difference between the sexes, and certain important changes in family structure.

For Additional Review

Compare the land adaptations of plants and of animals, including how each deals with taking in nutrients and water and with excreting wastes.

Multiple-Choice Questions

1. Which of the following groups thrives in extreme heats and acid environments?
   (A) Bryophytes
   (B) Protists
   (C) Archaebacteria
   (D) Fungi
   (E) Deuterostomia

2. Which of the following groups is best characterized as being heterotrophic and eukaryotic and as possessing plasma membrane?
   (A) Plantae
   (B) Animalia
   (C) Fungi
   (D) Viruses
   (E) Monera

3. A biologist captures an aquatic organism from a freshwater pond. It has a segmented body and a brainlike pair of central ganglia. It is also a bottom dweller, burrowing into the pond floor. Most likely this organism is a
   (A) protist
   (B) annelid
   (C) arachnid
   (D) eurypterid
   (E) reptile

4. In which of the following pairs are the organisms most closely related taxonomically?
   (A) Flatworms; rotifers
   (B) Mushrooms; tulips
   (C) Clams; lobsters
   (D) Sponges; hydras
   (E) Lancelets; roundworms

Directions: The group of questions below consists of five lettered headings followed by a list of numbered phrases or sentences. For each numbered phrase or sentence select the one heading that is most closely related to it and fill in the corresponding oval on the answer sheet. Each heading may be used once, more than once, or not at all.

Questions 5–9
(A) Chordata
(B) Arthropoda
(C) Annelida
(D) Platyhelminthes
(E) Porifera

5. Parasitic, lack a body cavity, acoelomates, freshwater and damp terrestrial habitats

6. Bodies segmented internally and externally, freshwater habitat, respiration occurs through the skin

7. Possess a notochord, a dorsal hollow nerve chord, pharyngeal slits, and a tail at some point during development

8. Segmented coelomates that have exoskeletons and jointed appendages

9. Sessile, possessing no nerves or muscles, have a central cavity called a spongocoel, are filter feeders

10. Systematicists categorize all living creatures into what three domains?
    (A) Bacteria, Euglena, Eukarya
    (B) Eubacteria, Archae, Eukarya
    (C) Archae, Plantae, Eukarya
    (D) Protista, Plantae, Eukarya
    (E) Prokaryota, Eukarya, Plantae

11. The cell walls of bacteria contain which of the following materials?
    (A) Peptidoglycans
    (B) Polynucleotides
    (C) Pyrimidines
    (D) Disaccharides
    (E) Ribose

12. The most common method of locomotion in prokaryotes occurs via the movement of
    (A) claws
    (B) the vascular cavity
    (C) flagella
    (D) tentacles
    (E) repeated alteration of cell shape

13. Heterotrophic cells are unable to synthesize which of the following compounds?
    (A) Cholesterol
    (B) Glycerol
    (C) Polypeptides
    (D) Ribonucleic acids
    (E) Glucose

14. In which of the following forms of parasitism do both organisms benefit?
    (A) Parasitic
    (B) Commensalistic
    (C) Mutualistic
    (D) Obligate
    (E) Heterotrophic

15. One example of a commensalistic symbiotic relationship is
    (A) the legume plant, which houses prokaryotes that fix nitrogen
    (B) the intestine of a pig, which houses roundworms
    (C) the vagina of a human, which hosts fermenting bacteria that produce acids
    (D) the intestine of a human, which houses *E. coli*
    (E) animal cells hosting a virus

16. Which of the following constitutes the kingdom containing the widest array of types of organisms?
    (A) Animalia
    (B) Plantae
    (C) Fungi
    (D) Protista
    (E) Bacteria

17. From an evolution perspective, certain of the organelles contained in animal cells are the result of what process?
    (A) Endosymbiosis
    (B) Conjugation
    (C) Transformation
    (D) Ectosymbiosis
    (E) Compartmentalization

18. Slime molds are contained in which of the following kingdoms?
    (A) Plantae
    (B) Archae
    (C) Eukarya
    (D) Protista
    (E) Euglena

Directions: The group of questions below consists of five lettered headings followed by a list of numbered phrases or sentences. For each numbered phrase or sentence select the one heading that is most closely related to it and fill in the corresponding oval on the answer sheet. Each heading may be used once, more than once, or not at all.

Questions 19–22
  (A) Bryophytes
  (B) Pteridophytes
  (C) Gymnosperms
  (D) Angiosperms
  (E) Fungi

19. Contain the mosses, lack vascular tissue

20. Contain the flowering plants, possess vascular tissue

21. Contain the ferns, are the seedless plants, and have vascular tissue

22. Contain the conifers, have seeds and vascular tissue

23. In alternation of generations in land plants, which of the following represents the haploid stage?
  (A) Zygote
  (B) Gametophyte
  (C) Sporophyte
  (D) Spore
  (E) Sporangia

24. All of the following are adaptations of land plants EXCEPT
  (A) stomata
  (B) cuticle
  (C) xylem
  (D) phloem
  (E) photosynthesis

25. Which of the following is one of the three types of bryophytes?
  (A) Red algae
  (B) Brown algae
  (C) Green algae
  (D) Liverworts
  (E) Lycophyta

26. Which of the following is the dominant stage of the life cycle for seedless vascular plants?
  (A) Gametophyte
  (B) Sporophyte
  (C) Zygote
  (D) Heterosporous
  (E) Homosporous

27. A student is collecting samples of plants from a field in Connecticut. He picks one from the ground to study it more closely. In the process he notices that the leaves show veins that run parallel from end to end. This plant is mostly likely a
  (A) eudicot
  (B) conifer
  (C) monocot
  (D) dicot
  (E) tracheid

28. The tiny filaments that comprise the body of fungi are known as
  (A) Hyphae
  (B) Mycelium
  (C) Chitin
  (D) Mycorrhizae
  (E) Basidium

Directions: The group of questions below consists of five lettered headings followed by a list of numbered phrases or sentences. For each numbered phrase or sentence select the one heading that is most closely related to it and fill in the corresponding oval on the answer sheet. Each heading may be used once, more than once, or not at all.

Questions 29–33
  (A) Mold
  (B) Yeast
  (C) Lichen
  (D) Mycorrhizae
  (E) Sac fungi

29. Mutualistic associations of plant roots and fungi

30. Unicellular fungi that live in damp environments

31. Symbiotic associations of photosynthetic microorganisms in a network of fungal hyphae

32. Phylum Ascomycota, contains many algae in associations with lichens

33. Rapidly growing, asexually reproducing fungi

34. All of the following terms reflect characteristics of all animals EXCEPT
  (A) multicellularity
  (B) heterotrophic nutrition
  (C) possessing nervous and muscle tissues
  (D) eukaryotic
  (E) diurnal

35. The body plans of sea anemones exhibit
  (A) bilateral symmetry
  (B) radial symmetry
  (C) dorsal and ventral symmetry
  (D) anterior and posterior symmetry
  (E) no symmetry

36. Most of the animal phyla originated in what geologic timespan?
  (A) The Paleozoic explosion
  (B) The Jurassic explosion
  (C) The Precambrian explosion
  (D) The Cambrian explosion
  (E) The Mesozoic explosion

37. The above represents which of the following forms of the cnidarian?
  (A) Medusa
  (B) Polyp
  (C) Hermaphrodite
  (D) Radula
  (E) Mantle

38. A body cavity that is not completely lined by tissue derived from the mesoderm is called a(n)
  (A) coelomate
  (B) subcoelomate
  (C) pseudocoelomate
  (D) anticoelomate
  (E) transcoelomate

39. Evolution of which feature or features enabled vertebrates to reproduce successfully on land?
    (A) The amniotic egg
    (B) Quadruped locomotion
    (C) Body hair
    (D) Opposable thumbs
    (E) Internal fertilization

40. Which of the following animals is characterized by a relatively short period of pregnancy followed by a period of nursing as it's offspring completes development?
    (A) Placental mammals
    (B) Marsupial mammals
    (C) Monotremes
    (D) Therapsids
    (E) Carnites

Free-Response Question

1. *Systematicists are scientists who study evolutionary relationships between organisms; they use scientific evidence to construct phylogenies that show these relationships. Recently systematicists have replaced the five-kingdom system with a more accurate, three-domain system, containing the Archae, the Eubacteria, and the Eukarya.*

▌ Describe how this new scheme for classification differs from the old one.
▌ Describe three types of evidence that scientists used to develop this three-domain system.

ANSWERS AND EXPLANATIONS

Multiple-Choice Questions

▌ 1. **(C) is correct.** The prokaryotes are split into two domains, Archae and Eubacteria. The Archae are known as extremophiles; they are often found in extreme (i.e. acidic or very hot) environments, such as the geysers of Yellowstone Park.

▌ 2. **(B) is correct.** Animals are heterotrophic—they are not capable of fixing carbon and must obtain it from phototropic organisms. They belong to the kingdom Eukarya, and they possess plasma membranes made up of phospholipids and proteins. This enables them to control the movement of substances into and out of the cell.

▌ 3. **(B) is correct.** Annelids are characterized by a body that is segmented internally and externally; they are 1 mm to 3 m in length and live in freshwater habitats, in the soil, and in the sea. Earthworms are a common annelid.

▌ 4. **(A) is correct.** The two organisms most closely related are the flatworms and rotifers. These two are in the category bilateria, protostomia, lophotrochozoa. The other organisms listed are not in the same categories, and all are in different phyla. However, they are all in the kingdom Animalia.

▌ 5. **(D) is correct.** Platyhelminthes are flatworms—acoelomates that have gastrovascular cavities.

▌ 6. **(C) is correct.** Phylum Annelida contains organisms with segmented body plans that inhabit freshwater or damp terrestrial habitats. They have closed circulatory systems, and respiration occurs through the skin.

7. (A) is correct. The chordates are mostly vertebrates, although there are two groups of invertebrate chordates as well. The chordates are grouped according to the presence of a notochord, a dorsal hollow nerve cord, pharyngeal slits, and a post-anal tail. Many of these features exist only during development.

8. (B) is correct. Including the lobsters, spiders, and other related organisms, the arthropods are characterized by having an exoskeleton (which is sometimes shed in a process called molting), jointed appendages, and segmentation.

9. (E) is correct. Porifera are the simplest invertebrates. They are sessile and have no muscles or nerve cells. They resemble a sac with holes in it, and they draw water into their central cavity (spongocoel), filtering out food particles.

10. (B) is correct. The three domains into which all the living organisms are placed by systematicists are Eubacteria, Archae, and Eukarya. Domains are one taxonomic level above that of kingdoms. Prokaryotes make up Archae and Eubacteria, whereas eukaryotes make up Eukarya.

11. (A) is correct. Peptidoglycans are polymers of modified sugars that are linked by short polypeptides that are different from species to species. This differs from the cellulose contained in plant cell walls.

12. (C) is correct. The flagella accounts for most movements in prokaryotes. Prokaryotic flagella are much smaller than eukaryotic flagella.

13. (E) is correct. Heterotrophs need at least one organic compound in order to make others. Chemotrophs get their energy from chemicals in the environment. Autotrophs can make organic compounds from CO_2.

14. (C) is correct. In mutualistic symbiosis, both organisms benefit from the association; in commensalistic symbiosis, one organism benefits while the other is neither helped nor harmed; and in parasitic symbiosis, one organism benefits while the other is harmed.

15. (D) is correct. The *E. coli* living in the human intestine constitutes commensalistic symbiosis. In this situation, the *E. coli* benefit from the nutrients passing through the intestine, while the human is usually neither helped nor harmed by their presence.

16. (D) is correct. Protists are by far the most varied kingdom. Most of the protists are unicellular, but there are multicellular protists, too. Protists can be divided into protozoa (animal-like protists) and algae (plant-like protists).

17. (A) is correct. Endosymbiosis—the theory of serial endosymbiosis—proposes that mitochondria and chloroplasts used to be small prokaryotes that lived in larger cells. Eventually they became a permanent functional part of the cell.

18. (D) is correct. Slime molds are decomposers and are part of the kingdom Protista. They have structures called pseudopodia that are used for moving and feeding.

19. (A) is correct. Bryophytes are one of the four main types of land plants. The most common bryophytes are the mosses, which do not have vascular tissue and are relatively simple.

20. (D) is correct. The angiosperms are the flowering plants we see around us. They have vascular tissue and the flower is their reproductive structure. They also have seeds enclosed within their flower.

21. (B) is correct. Pteridophytes are seedless plants that contain vascular tissue. They are not as highly evolved as gymnosperms and angiosperms, and one prominent member of this group is the ferns.

22. (C) is correct. Gymnosperms are seed plants that have "naked" seeds. These seeds are not enclosed in any specialized chambers. But these plants do possess vascular tissue. Pine trees are a common example of gymnosperms.

23. (B) is correct. The cells of the gametophyte are haploid ($1n$) and have a single set of chromosomes. The fusion of sperm and egg cells during fertilization produces the zygote, which is diploid. The zygote divides mitotically to produce the sporophyte.

24. (E) is correct. Photosynthesis in plants does not necessarily contribute to their ability to conserve water or to store water in their structure. The other plant structures listed are all evolutionary adaptations for terrestrial plants.

25. (D) is correct. There are three phyla of bryophytes—mosses, liverworts, and hornworts. In bryophytes, the gametophyte is the dominant generation, whereas sporopytes are typically present only briefly.

26. (B) is correct. The sporophyte stage is the dominant one in the life cycle of the seedless vascular plant. The sporophyte is the diploid stage, whereas gametophytes are tiny and exist only briefly.

27. (C) is correct. This plant is probably a monocot. Monocots have veins that run parallel, whereas dicots have veins that have netlike vein arrangement in their leaves. Grass blades are a good example of monocots.

28. (A) is correct. Hyphae are filaments made up of tube-like walls that surround cytoplasm and plasma membranes. They group together to form a woven mat called a mycelium.

29. (D) is correct. Mycorrhizae are mutual associations of fungi and plant roots. They exchange minerals extracted from the soil and nutrients produced by the plants.

30. (B) is correct. Yeasts are unicellular fungi that live in moist environments or liquids. They reproduce by cell division or budding.

31. (C) is correct. Lichens are symbiotic associations of millions of photosynthetic organisms in a network of fungal hyphae. They are very hardy and frequently colonize newly broken rock faces.

32. (E) is correct. Sac fungi are in the phylum Ascomycota, and they live in all sorts of environments—marine, freshwater, and terrestrial. They produce small spores called asci, thus their name.

33. (A) is correct. Molds are rapidly growing fungi that reproduce asexually. They grow on a variety of mediums and go through a series of different reproductive stages.

34. (E) is correct. The first four answer choices represent characteristics common to all animals; the last one does not. The quality of being diurnal (the opposite of nocturnal—being active during the day) is not a requirement for belonging to the kingdom Animalia.

35. (B) is correct. Sea anemones can best be described as having radial symmetry. Radial animals have a top and bottom but no head or rear end, or left and right.

■ **36. (D) is correct.** The Cambrian explosion that occurred about 530 million years ago led to the relatively quick appearance of most of the major animal phyla. Many fossils dating from that era have been discovered.

■ **37. (A) is correct.** This represents the medusa form of the cnidarian. Despite the fact that they can move, polyps (like hydras and sea anemones) adhere to a substrate with their mouth/tentacle side up. Medusas (jellyfish) are free swimming and "upside down" compared to polyps—that is, their oral/tentacle side faces downward. Cnidarians are characterized by having radial symmetry, a gastrovascular cavity, and cnidocytes—cells that function in defense and capture of prey.

■ **38. (C) is correct.** Animals that have no cavity between their digestive tract and their body wall are acoelomates. Those that have this are pseudocoelomates, and those that lack a true coelom are coelomates.

■ **39. (A) is correct.** The amniotic egg is composed of extraembryonic membranes that function in gas exchange, waste storage, and the delivery of nutrients to the embryo.

■ **40. (B) is correct.** Marsupials are predominantly located in Australia, where they have radiated to fill niches occupied by placental mammals elsewhere. Marsupials have a short period of pregnancy, and the embryo completes its development nursing within the pouch of the mother.

Free-Response Question

1. The old five-kingdom system contained the kingdoms Monera, Protista, Plantae, Fungi, and Animalia. However, scientists recently determined that the Monera kingdom should be further separated into two groups. Thus, they split Monera into Archae and Eubacteria. They realized that the Archae and Eubacteria are as different from each other as they are different from us, the Eukarya. The new system is a three-domain system, with the domains Archae, Eubacteria, and Eukarya. Archae and Eubacteria are composed of prokaryotes, and Eukarya is made up of eukaryotes. From the old five-kingdom system, the Protista, Plantae, Fungi, and Animalia are all in the domain Eukarya.

Three methods or types of evidence that scientists use to classify organisms and study their degree of evolutionary relatedness are fossil evidence, the study the structure and development of organisms, and molecular evidence.

Fossils are the remains or impressions of organisms from the past, and the fossil record is the order in which fossils are uncovered, layer by layer, in the earth. The type of rock that fossils are most frequently found in is sedimentary rock—this is because sedimentary rock is formed when sand and silt settle to the bottom of lakes, ponds, oceans, and rivers and then deposits pile on top of them, compressing them into solid rock. Sedimentary rock also forms strata; layers that represent certain geologic periods; this is what creates the fossil record. Most frequently the parts of an organism that appear in fossils are the calcified shells of invertebrates and some bones of vertebrates. Systematists use relative dating—basically just correlating finds and data between sites to determine relative ages, as well as radiometric dating (the measurement of radioactive isotopes with known half-lives, in rocks) to determine the ages of fossils.

Systematists can also study patterns of homologous structures to determine evolutionary relationships. Species that were derived from the same ancestor should have similarities, and these similarities are called homologies. Here, scientists would look for homologous structures—structures that are similar in different species—and use these to tie organisms together.

The third way that systematists can study evolutionary relationships is on a molecular level. Since DNA is passed on from generation to generation, related species should share common genes; and the more recently the species branched off from a common ancestor, the more similar their DNA should be. Studying the DNA of organisms makes it possible for systematists to determine the degree of evolutionary difference between two species that are nearly identical in appearance, and it also allows systematists to judge the relatedness of two species that they might not guess would be related at all, based on external appearance. It is a much more precise and quantitative method for appraising evolutionary relatedness.

This response uses the following key terms in context, showing the writer's knowledge of their meanings and relatedness:

Monera	*Animalia*	*systematists*
Protista	*Archae*	*homologous structures*
Plantae	*Eubacteria*	*DNA*
Fungi	*Eukarya*	

This response also accurately describes the rationale behind the process of the reorganizing of the five-kingdom system into the three-domain system. It goes on clearly to describe three methods for studying the evolutionary relatedness of species, thus answering the question in a complete and organized way. The response could be improved with a closing statement that ties the evidence cited back to the original question about domains. Test Tip: Always check your final response to make sure it addresses the original question.

CHAPTER 6

Plant Form and Function

Plant Structure and Growth: The Plant Body

▌ Plants have a **root system** beneath the ground and a **shoot system** above the ground.

▌ There are two types of roots: **fibrous roots**, which are made up of a mat of thin roots that are spread just below the soil's surface, and **taproot** systems, which are made up of one thick, vertical root with many lateral roots that come out from it.

▌ Stems are made up of **nodes** (points at which leaves are attached) and **internodes** (the parts of the stem between the nodes).

▌ **Axillary buds** are located in the V formed between the node and the stem, and these have the potential to form a branch.

▌ The **terminal bud** is located at the top end of the stem and is where growth usually occurs. In **apical dominance**, the terminal bud prohibits the growth of the axillary buds.

▌ Leaves are the main sites of photosynthesis in plants. Leaves consist of a **blade**, which is flat, and the **petiole**, which joins the leaf to a node of the stem.

▌ The leaf, stem, and root of plants are all considered organs, and all plant organs are composed of three tissue types.

1. **Dermal tissue** is a single layer of closely packed cells that covers the entire plant and protects it.

2. **Vascular tissue** is continuous through the plant and responsible for transporting materials between the roots and shoots. Vascular tissue is made up of:

    (a) **xylem**, which transports water and minerals up from the roots

    (b) **phloem**, which transports food from the leaves to the other parts of the plant

    (c) **tracheids** and **vessel elements**, which are components of xylem, are dead cells that form a conduit through which water passes

    (d) in phloem, tubes formed by chains of cells are called **sieve-tube members**.

3. **Ground tissue** is defined as anything that isn't dermal tissue or vascular tissue. Its divided into pith (inside the ring of the ground tissue) and cortex (outside of the ring of ground tissue).

▌ Plants have three cell types.

    1. **Parenchyma cells** perform most of the metabolism (including photosynthesis) in the cell and are present throughout the plant.

2. **Collenchyma cells** are grouped in cylinders, and help support growing parts of the plant.
3. **Sclerenchyma cells** exist in parts of the cell that are no longer growing, and they have tough cell walls. Two types of these are specialized just for support—**fibers** and **sclereids**.

The Process of Plant Growth and Development

▌ Plants are called **annuals** if their life span occurs over one year, **biennials** if their life occurs over the course of two years, **and perennials** if their lives occur over a number of years.

▌ **Meristems** are embryonic growths that exist in a plant's growing regions.

▌ Apical meristems are located at the tips of roots and in buds of shoots, and these are the sites of cell division.

▌ **Primary growth** occurs when the plant grows at the apical meristems, while **secondary growth** (which occurs in woody plants) is when the shoots and roots of a plant thicken. Secondary growth is the product of **lateral meristems**.

▌ The root cap covers the tip of the root in a plant, and contains three types of cells in various stages of growth, in the following zones:

1. The **zone of cell division** includes the apical and primary meristems, where cells divide rapidly, and contains a quiescent center, where cells divide much more slowly. Above the apical meristem, the newly divided cells form three concentric cylinders of cells that continue to divide—these are the **protoderm**, the **procambrium**, and **ground meristem**, which eventually produce the three main tissues.
2. Above the zone of cell division is the **zone of elongation**, in which cells elongate significantly.
3. In the **zone of maturation**, the three systems in primary growth complete their differentiation and become functionally mature.

▌ In roots, the procambrium gives rise to the **stele**, a central cylinder of vascular tissue in which xylem and phloem both develop. The outermost layer of the stele is the **pericycle**.

▌ The **endodermis** is a ring, one cell thick, that separates the cortex from the stele.

▌ At a shoot, the apical meristem is a dome of dividing cells at the tip of a terminal bud. The apical meristem gives rise to protoderm, procambrium, and ground meristem, which also develop into the vascular tissue systems.

▌ The epidermis of the underside of the leaf is interrupted by stomata, which are small pores flanked by guard cells, which open and close the stomata.

▌ In leaves, the ground tissue is sandwiched between the upper and lower epidermis, in the mesophyll. It is made up of parenchyma cells, the sites of photosynthesis.

▌ Two lateral meristems take part in plant growth: the **vascular cambium**, which produces secondary xylem (wood), and the **cork cambium**, which produces a tough covering that replaces epidermis.

- Early in secondary growth, the epidermis dries up and is replaced by cork cambium. Cork cambium produces **cork cells**, and together the cork plus cork cambium make up the **periderm**.
- Openings, or splits in the cork cambium, called lenticels, enable the cells within the trunk to exchange gases with the outside, and continue cell respiration.
- **Bark** is all the tissues outside the vascular cambium.

Mechanisms of Plant Growth and Development

Growth in plants occurs by cell elongation and expansion, as well as by cell division.

Transport in Plants: An Overview of Transport Mechanisms in Plants

- Three kinds of transport occurs in plants:

    1. The uptake and loss of water and minerals from individual cells, for instance in a root cell or leaf cell
    2. Transport of substances short distances, from cell to cell
    3. Transport of sap within the xylem and phloem, throughout the entire plant

- In plants, the uptake of water across cell membranes occurs through osmosis, the passive transport of water across a membrane.
- The **water potential** is defined as the combined effects of solute concentration and the pressure that the cell wall contributes.
- **Turgor pressure** is the pressure exerted against the cell wall when the cell is filled with water.
- **Aquaporins** are the channels in the plant cell walls specifically designed for the passage of water.
- Plants have a **tonoplast** surrounding their vacuoles; the tonoplast regulates molecules going into and out of the vacuole.
- The **symplast** is a continuum of cytoplasm that is connected by plasmodesmata between cells.
- The **apoplast** is the nonliving continuum that is formed by the extracellular pathway formed by the continuous matrix of cell walls.
- Water flows through both the apoplast and the symplast.
- **Bulk flow** is the movement of water through the plant by pressure.

Absorption of Water and Minerals by Roots

- In the roots, most water absorption occurs near the tips, in the root hairs.
- **Mycorrhizae** are roots associated with symbiotic fungi. These absorb water and certain minerals.
- Water and minerals in the root cortex must pass through the **endodermis** in order to enter the rest of the plant. The endodermis surrounds the stele.
- The endodermis contains the **Casparian strip**, which prevents substances from going around the cells. Water and minerals must therefore pass through an endodermal cell to enter the vascular tissue.

Transport of Xylem Sap

▎ **Transpiration** is the loss of water vapor from the leaves and other parts of the plant in contact with air.

▎ There are two ways in which water is pulled up through the plant.

1. **Root pressure** occurs when water flowing in from the root cortex generates a positive pressure that forces fluid up through the xylem.
2. The **transpiration-cohesion-tension mechanism**—Water is lost through transpiration at the leaves of the plant, and this creates a negative pressure, which draws water up through the plant. The cohesion of water due to hydrogen bonding enables it to form a column, which is drawn up through the xylem with the help of adhesion (the fact that water molecules are attracted to the plant cell walls).

The Control of Transpiration

▎ **Guard cells** control the size of the opening in the stomata and therefore regulate the plants water intake.

▎ Guard cells control the size of the stomata opening by changing shape, widening or closing up the gap between them. They increase in size by taking up water. This causes them to swell and seal off the stomata opening.

▎ Three factors influencing the opening and closing of stomata are light, depletion of CO_2 in air spaces in the leaf, and the internal clock of the guard cells.

Translocation of Phloem Sap

▎ Phloem transports organic products of photosynthesis from the leaves throughout the plant.

▎ Sieve tubes carry food from a **sugar source** (a plant organ where sugar is being produced) to a **sugar sink** (an organ that consumes sugar).

▎ Flow through the phloem occurs mainly as a result of bulk flow—loading of sugar into cells creates a high solute concentration at the source end of the sieve tube, and this lowers the water potential and causes the water to flow in the tube.

Plant Nutrition: Nutritional Requirements of Plants

▎ Mineral nutrients are those required by plants that are chemical elements absorbed from the soil as inorganic ions.

▎ Plants need nine **macronutrients** in great amounts: carbon, oxygen, hydrogen, nitrogen, sulfur, phosphorus, potassium, calcium, and magnesium. They need at least eight **micronutrients** in small amounts: chlorine, iron, boron, manganese, zinc, copper, molybdenum, and nickel. Essential nutrients are those that the plant needs to complete the life cycle.

The Role of Soil in Plant Nutrition

Topsoil is defined as a mixture of particles from rock, living organisms, and **humus** (partially decayed organic material).

The Special Case of Nitrogen as a Plant Nutrient

▌ In order for plants to absorb nitrogen, it must first be converted to NH_4^+ or NO_3^-.

▌ The main source of nitrogen for plants is the decomposition of humus by microbes.

▌ **Nitrogen-fixing bacteria** convert N_2 to NH_3, in the process of **nitrogen fixation**. This is a form of nitrogen that plants are able to use.

▌ Plant roots, specifically legume roots, have swellings called **nodules** that are composed of plant cells that contain nitrogen-fixing bacteria. This is a mutualistic symbiotic association.

Nutritional Adaptations: Parasitism and Predation by Plants

▌ Mistletoe is a **parasitic plant**. Some parasites are not photosynthetic and rely on other plants for their nutrients.

▌ **Epiphytes** are not parasitic, they just grow on the surfaces of other plants.

▌ **Carnivorous plants** are photosynthetic, but they get some nitrogen and other minerals by digesting small animals.

Plant Reproduction and Biotechnology: Sexual Reproduction

▌ Angiosperms and other plants undergo an alternation of generations in their life cycle, in which the diploid plant or sporophyte, meiotically produces haploid spores which then develop into male and female haploid plants. These haploid plants, gametophytes, develop and produce gametes, which undergo fertilization to form new sporophytes.

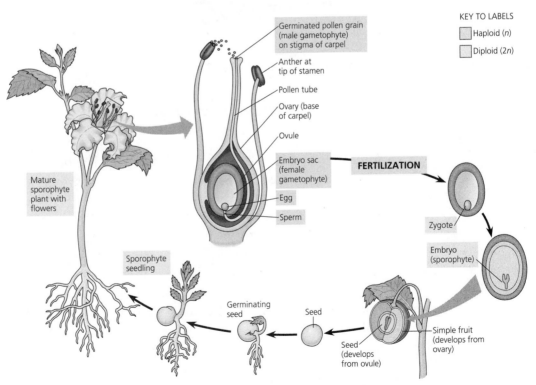

Simplified Overview of Angiosperm Life Cycle

- In angiosperms, the sporophyte is the dominant generation, because that is the form of the plant we see.
- Flowers are the reproductive organs of angiosperms. Some important flower structures are:

    - **Sepals**—which protect the floral bud before it opens
    - **Petals**—which attract insects and other pollinators to the plant with its color and fragrance
    - **Stamens**—male reproductive organs
    - **Carpels**—female reproductive organs

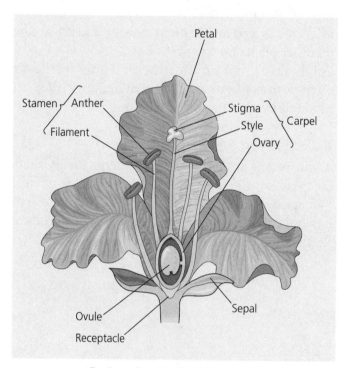

Review of an Idealized Flower

- **Monoecious** species have staminate and carpelate flowers located on the same plant, whereas **dioecious** species have staminate flowers and carpelate flowers on different plants.
- In the sporangia of an anther there are many diploid cells called microsporocytes. Each microsporocyte undergoes meiosis to produce four haploid microspores, which can eventually become haploid male gametophytes.
- In the ovary, **ovules** form—each of them containing a single sporangium. Within the sporangium, a single cell called the **megasporocyte** grows and undergoes meiosis to produce four haploid megaspores. In many angiosperms, only one of the megaspores survives, and a series of steps produces an ovule.

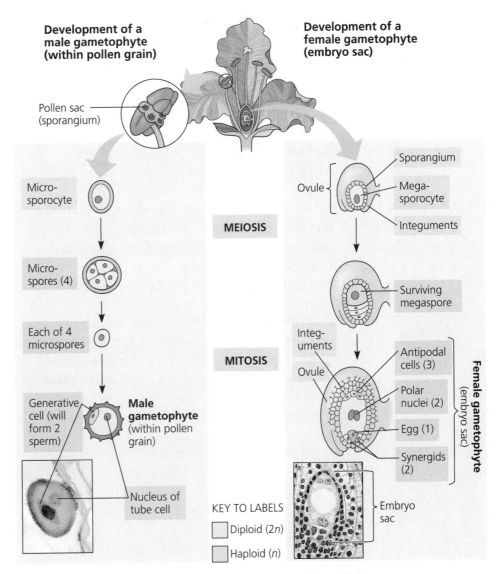

Development of a male gametophyte (within pollen grain)

Development of a female gametophyte (embryo sac)

Pollen sac (sporangium)

Micro-sporocyte

Micro-spores (4)

Each of 4 microspores

Generative cell (will form 2 sperm)

Male gametophyte (within pollen grain)

Nucleus of tube cell

MEIOSIS

MITOSIS

Ovule

Sporangium

Mega-sporocyte

Integuments

Surviving megaspore

Integuments

Ovule

Antipodal cells (3)

Polar nuclei (2)

Egg (1)

Synergids (2)

Female gametophyte (embryo sac)

KEY TO LABELS

☐ Diploid (2n)

☐ Haploid (n)

Embryo sac

The Development of Angiosperm Gametophytes

▌ While some flowers self-fertilize, others have methods by which self-fertilizing is prevented, in order to maximize genetic variety. One of these is **self-incompatibility**, in which a plant can reject the pollen of itself or a closely related individual.

▌ When a pollen grain lands on a stigma, it germinates and produces a **pollen tube** that extends toward the ovary. Then, one sperm fertilizes the egg to form the 2n embryo. The other combines with the two polar nuclei to form a triploid nucleus (3n), which will eventually give rise to the **endosperm**. The **endosperm** will nourish the plant embryo. This process of forming an embryo and endosperm is called **double fertilization**.

▌ After double fertilization, the ovule develops into a seed, and the ovary develops into a **fruit**, which encloses the seed.

▌ As the seed matures, it enters **dormancy**, in which it has a low metabolic rate and its growth and development are suspended.

The seed resumes growth when there are suitable environmental conditions, or when it is located at a place suitable for plant development.

Asexual Reproduction

Asexual reproduction produces clones. Asexual reproduction is also known as vegetative reproduction. In this process, fragmentation occurs, in which pieces of the parent plant break off to form new individuals who are exact genetic replicas of the parent.

Many plants are capable of both asexual and sexual reproduction.

Plant Responses to Internal and External Signals: Plant Responses to Hormones

Hormones are defined as chemical messengers that act to coordinate the different parts of an organism. They are produced by one part of the body and transported to another.

A **tropism** is a growth response in a plant that results in the plant growing either toward or away from a stimulus.

Phototropism is the growth of a shoot in a certain direction as a response to light. **Positive phototropism** is the growth of a plant toward light; **negative phototropism** is growth of the plant away from light.

Plants grow toward light as a result of the plant hormone auxin moving down from the apex to enter the cell that are receiving less light; this stimulates them to elongate.

An Overview of Plant Hormones

Hormone	Where Produced or Found in Plant	Major Functions
Auxin (IAA)	Embryo of seed, meristems of apical buds, young leaves	Stimulates stem elongation (low concentration only) root growth, cell differentiation, and branching; regulates development of fruit; enhances apical dominance; functions in phototropism and gravitropism.
Cytokinins (such as zeatin)	Synthesized in roots and transported to other organs	Affect root growth and differentiation; stimulate cell division and growth; stimulate germination; delay senescence.
Gibberellins (such as GA₃)	Meristems of apical buds and roots, young leaves, embryo	Promote seed and bud germination, stem elongation, and leaf growth; stimulate flowering and development of fruit; affect root growth and differentiation
Abscisic acid	Leaves, stems, roots, green fruit	Inhibits growth; closes stomata during water stress; counteracts breaking of dormancy
Ethylene	Tissues of ripening fruits, nodes of stems, aging leaves and flowers	Promotes fruit ripening, opposes some auxin effects; promotes or inhibits growth and development of roots, leaves, and flowers, depending on species
Brassinosteroids (such as brassinolide)	Seeds, fruits, shoots, leaves, and floral buds	Inhibits root growth; retards leaf abscission; promotes xylem differentiation

Plant Responses to Light

▌ **Photomorphogenesis** is the term used to describe the effects of light on plant morphology.

▌ Blue light has the greatest effect on plant growth and movement. Plants use three different pigments to detect blue light: **cryptochromes**, **phototropin**, and **zeaxanthin**.

▌ **Phytochromes** are pigments that are involved in many of a plants responses to light.

- Many of the plant processes occur more and less frequently depending on the time of day, in response to changes in light, temperature, and humidity.
- **Circadian rhythms** are physiological cycles that have a frequency of about 24 hours and that are not paced by a known environmental variable.
- A physiological response to a photoperiod, like flowering, is called **photoperiodism**.
- **Short-day plants** require a period of light shorter than a certain critical length in order to flower. **Long day plants** flower in the late spring or early summer—they require the most daylight to flower. **Day neutral plants** can flower in days of any length.
- It is night length—not day length—that controls flowering and certain other responses to photoperiod.

Plant Responses to Environmental Stimuli Other than Light

- **Gravitropism** is a plant's response to gravity. Roots show positive gravitropism, and grow toward the ground, whereas shoots show negative gravitropism and grow toward the sky.
- **Thigmomorphogenesis** is the change in form of a plant that results from mechanical disturbance.
- **Thigmotropism** is directional growth in a plant as a response to a touch.
- In times of drought, the guard cells lose **turgor**. This causes the stomata to close; young leaves will stop growing, and they will roll into a shape that prevents transpiration from occurring as quickly. Also, deep roots continue to grow, while those near the surface (where there isn't much water) do not grow very quickly.
- In times of flooding, certain cells in the root cortex die, which creates air tubes in the cortex that enable the plant cell to continue cell respiration.
- In heat stress environments, plants produce **heat-shock proteins**, which perform an as-yet unknown function in the cell.
- In cold stress situations, plants respond by altering the composition of their cell membranes.

Plant Defense: Responses to Herbivores and Pathogens

- Some physical defenses plants have against predators are thorns, chemicals such as distasteful or poisonous compounds, and airborne attractants that bring other animals to kill the herbivores.
- The first line of defense against viruses for a plant (as for humans) is the epidermal layer.
- The plant is capable of recognizing plant pathogens and dealing with them in complex biochemical ways.

For Additional Review:

Make a list of the elements of plant structure that make it ideally suited for trapping and processing the sun's energy, and for the process of carbon fixation.

Multiple-Choice Questions

1. Which of the following is primarily responsible for fruit ripening in plants?
   (A) Ethylene
   (B) Auxin
   (C) Gibberellins
   (D) Abscissic acid
   (E) Brassinosteroids

2. Which of the following processes is responsible for the bending of the stem of a plant toward a light source?
   (A) The amount of chlorophyll produced on the side facing the light increases.
   (B) Cell division on the side facing the light increases in speed.
   (C) Cell division on the side away from the light increases in speed.
   (D) The cells on the side of the stem facing the light elongate.
   (E) The cells on the side of the stem away from the light elongate.

3. The driving force for the movement of materials in the xylem of plants is
   (A) gravity
   (B) root pressure
   (C) transpiration
   (D) the difference in osmotic pressure between the source and the sink
   (E) osmosis

4. The loss of leaves that some plants experience due to the onset of autumn is a result of which hormone?
   (A) Auxin
   (B) Gibberellic acid
   (C) Cytokinin
   (D) Ethylene
   (E) Abscisic acid

5. In plants that exhibit an alternation of generations in their life cycle, which of the following is the reason why the sporophyte stage is said to be the dominant phase?
   (A) The gametophyte stage lasts longer than the sporophyte stage.
   (B) The sporophyte stage lasts longer than the gametophyte stage.
   (C) The sporophyte is the form of the plant that is independent and conspicuous.
   (D) The gametophyte is the form of the plant that is independent and conspicuous.
   (E) The sporophyte bears a reproductive structure.

6. In plants, translocation occurs as a result of
   (A) a difference in water potential between a sugar source and a sugar sink
   (B) transpiration
   (C) cohesion-adhesion
   (D) active transport by sieve-tube members
   (E) active transport by tracheid and vessel cells

Directions: The group of questions below consists of five lettered headings followed by a list of numbered phrases or sentences. For each numbered phrase or sentence select the one heading that is most closely related to it and fill in the corresponding oval on the answer sheet. Each heading may be used once, more than once, or not at all.

Questions 7–11
(A) Abscissic acid
(B) Auxin
(C) Cytokinins
(D) Ethylene
(E) Gibberellins

7. Produced in the roots and affects root growth and differentiation

8. Produced in tissues of ripening fruits and affects leaf abscission

9. Produced in the meristems of buds and roots, promotes seed and bud germination

10. Produced in the leaves, stems, and roots, and inhibits growth and closes stomata during drought

11. Produced in seed embryo and apical meristems, stimulates stem elongation and root growth

12. Fibrous root systems consist of a network of thin roots that spread out beneath the surface of the soil. Which of the following is the correct name for a system of roots that grows in one large vertical root with smaller lateral off-shoots?
(A) Tuberoot
(B) Taproot
(C) Toproot
(D) Stabroot
(E) Bladeroot

13. The three types of plant tissue, in order from the outside of the plant to the inside of the plant, are
(A) vascular, ground, dermal
(B) vascular, dermal, ground
(C) ground, vascular, dermal
(D) ground, dermal, vascular
(E) dermal, ground, vascular

14. A plant whose life span occurs over the course of two years is known as a(n)
(A) annual
(B) diannual
(C) biannual
(D) perennial
(E) seasonal

15. The region of the plant in which the parenchyma cells are located that are involved in photosynthesis is called
(A) spongeophyll
(B) mesophyll
(C) epidermis
(D) xylem
(E) phloem

16. In a mesophyll cell of a leaf, the synthesis of ATP takes place in the mitochondria and which of the following other cell organelles?
(A) Chloroplasts
(B) Golgi apparatus
(C) Nucleus
(D) Ribosomes
(E) Lysosomes

17. All of the following enhance the uptake of water by a plant's roots EXCEPT
(A) root hairs
(B) the large surface area of cortical cells
(C) mycorrhizae
(D) the attraction of water and dissolved minerals to root hairs
(E) gravitational force

18. The barrier located in the epidermal wall, which prevents the passage of unwanted minerals into the vascular tissue, is called the
(A) Narnian strip
(B) Octavian strip
(C) Casperian strip
(D) Gotham strip
(E) Tanzanian strip

19. All of the following contribute to the closing of stomata during the day EXCEPT
(A) water deficiency
(B) wilting
(C) high temperatures
(D) excessive rainfall
(E) excessive transpiration

20. Which of the following constitute macronutrients, elements needed by the plant in relatively large amounts?
(A) Carbon, oxygen, nitrogen and hydrogen
(B) Carbon, oxygen, nitrogen, and chlorine
(C) Carbon, oxygen, nitrogen, and iron
(D) Carbon, oxygen, nitrogen, and zinc
(E) Carbon, oxygen, nitrogen, and copper

21. Which term describes the symbiotic relationship between the roots of legumes and fungi?
(A) Bacteroids
(B) Humus
(C) Casperian strip
(D) Apoplast
(E) Mycorrhizae

22. Which of the following terms describes a species of plant that has male and female reproductive systems on the same individual plant?
(A) Deciduous
(B) Monoecious
(C) Dioecious
(D) Dihybrid
(E) Monohybrid

23. The point of attachment of a plant leaf and stem is called the
(A) carpal
(B) petiole
(C) blade
(D) internode
(E) axillary attachment

24. Which vascular tissue in plants is responsible for carrying sugars down from the leaves to the rest of the plant?
(A) Xylem
(B) Phloem
(C) Dermal tissue
(D) Tracheids
(E) Vessel elements

25. Which of the following colors of visible light has the greatest effect on plants?
(A) Red
(B) Orange
(C) Yellow
(D) Green
(E) Blue

Free-Response Question

1. *Describe the following processes in the plant life cycle and why they are important for the plant to complete, listing all of the plant hormones involved and describing their function:*

   (a) elongation of the plant shoot;
   (b) the process by which plants orient themselves with respect to the sun;
   (c) and photoperiodism.

ANSWERS AND EXPLANATIONS

Multiple-Choice Questions

▌ **1. (A) is correct.** Ethylene is a plant hormone that causes fruit to ripen. It also causes apoptosis, or programmed cell death in plant cells; it changes patterns of plant growth as a response to mechanical stress; and it causes the loss of leaves when autumn comes.

▌ **2. (E) is correct.** When the light source on either side of a plant is uneven, the plant will grow toward the light source. This is the result of auxin moving from the apex down to the cells that are less exposed to light, and causing them to elongate faster than the cells on the side that is illuminated.

▌ **3. (C) is correct.** The driving force behind the movement of sap in xylem (in the direction from the roots to the leaves) is the transpiration of water through the stomates on the leaves. The mechanism responsible for movement up through the xylem is the transpiration-cohesion-tension mechanism, and it occurs through bulk flow, in which fluid moves because of a pressure difference at opposite ends of a tube. The pressure is created by transpiration from the leaves, and contributing to the movement of water and minerals up the plant are gradients of water potential from cell to cell within the plant.

▌ **4. (D) is correct.** The loss of leaves that some plants experience as a result of the onset of autumn is due to the plant hormone ethylene. The technical term for this loss of leaves is leaf abscission, and it occurs in order to prevent plants from dehydrating during the winter, when they can't take in water from the ground because it is frozen. Ethylene and auxin are both involved in this process, but it is the increase in ethylene that ultimately triggers abscission.

▌ **5. (C) is correct.** In vascular plants that undergo an alternation of generations, the dominant form of the plant is the sprorophyte—it is the full grown plant that we see growing in a field. It has a reproductive structure called a flower, which creates gametes, and these eventually fuse to form a sporophyte embryo, which develops into a full grown mature sporophyte.

▌ **6. (A) is correct.** In plants, phloem is responsible for carrying sugar made in the leaves to other locations that are incapable of photosynthesis. This process is called translocation. Phloem is made up of sieve-tube members that are arranged end to end in long sieve tubes. In phloem, sugar travels from a sugar source—any site in the plant involved in photosynthesis (though this is usually mature leaves)—to a sugar sink, which is any site in the plant not engaged in photosynthesis.

7. (C) is correct. Cytokinins are plant hormones that are involved in the control of cell division and differentiation, the control of apical dominance in plants, and have some anti-aging effects on plant tissues. Cytokinins often act in concert with auxin.

8. (D) is correct. Ethylene has the following effects on plants: It initiates a response to mechanical stress (like when a plant must grow around an object in its regular path of growth); it is involved in apoptosis, which is programmed cell death (which occurs when plants shed their leaves, for instance); it is involved in leaf abscission; and it is responsible for the ripening of fruit.

9. (E) is correct. Gibberellins are plant hormones that are responsible for many different effects in plants, but three main ones are the elongation of the plant stem (they stimulate both cell division and cell elongation), fruit growth, and germination, which is the process by which a seed breaks dormancy and begins to grow.

10. (A) is correct. Abscisic acid is a plant hormone that prevents the seed from immediately germinating—it is responsible for seed dormancy. This hormone is also responsible for closing the stomata on plant leaves in times of drought—when plants start to wilt, abscisic acid rushes to the leaves and closes the stomata to prevent further transpiration.

11. (B) is correct. Auxins appear to have many functions in all types of plants. They are involved in cell elongation, the formation of lateral and adventitious roots. They act as herbicides, and they are involved in phototropisms. They are also involved in secondary growth, and this also promotes the growth of fruit.

12. (B) is correct. There are two main root systems in plants, the taproot system and the fibrous root system. Monocots, including grasses, usually have the fibrous root system, which firmly anchors them into the ground, while many dicots have a taproot system, with one long thick root extending down and smaller lateral branch roots shooting off of it.

13. (E) is correct. The three types of tissue that make up plant organs, from the external-most layer to the innermost layer are dermal, ground, and vascular. The dermal layer is a single layer of very tightly packed cells that serves to protect the plant from harm. The vascular tissue consists of xylem and phloem, and it is responsible for transporting water, minerals, and food around the plant. The ground tissue has various functions, but it serves in neither transport nor protection.

14. (C) is correct. A plant whose life cycle spans two years is known as a biannual. One whose life cycle spans one year is known as an annual, and plants who live for many years are known as perennials.

15. (B) is correct. The name of the region of the plant leaf in which parenchyma cells are situated—and the site of photosynthesis in plants—is called mesophyll. The mesophyll lies between the upper and lower epidermis of the leaf, and consists mainly of parenchyma cells. Dicots have two regions of mesophyll, spongy mesophyll and palisade mesophyll.

16. (A) is correct. The production of ATP in plant cells occurs in the mitochondria, as it does in the cells of animals, but it also occurs in the chloroplasts during photosynthesis. The light reactions of photosynthesis convert solar energy to the chemical energy of ATP and NADPH, and these light reactions take place in the chloroplasts, in the mesophyll cells of the plant leaf.

17. **(E) is correct.** All of the factors listed aid in the uptake of water and minerals by the roots of a plant except the last choice, gravity. Water and minerals flow from the soil into the epidermis of the plant, then through the root cortex, and then into the xylem of the plant, to be transported throughout the plant body.

18. **(C) is correct.** The Casparian strip is a belt made of a waxy material that runs through all of the endodermal cells to create a ring, which protects the vascular tissue from unwanted minerals. The water and mineral solution must pass from the cortex through the endodermis, and because of the Casparian strip, in order to pass the endodermal wall, the solution must be screened through the plasma membrane of an endodermal cell.

19. **(D) is correct.** All of the answers listed, with the exception of excessive rainfall, are factors that would cause the stomata of a leaf to close during the day. All of the factor with the exception of answer *D* would cause the plant to dehydrate in one way or another.

20. **(A) is correct.** The macronutrients (elements required in large amounts) in plants are carbon, oxygen, hydrogen, nitrogen, sulfur, phosphorus, potassium calcium, and magnesium. The micronutrients (elements needed only in trace amounts) are chlorine, iron, boron, manganese, zinc, copper, molybdenum, and nickel.

21. **(E) is correct.** Mycorrhizae are mutualistic symbiotic associations of roots and fungi. The fungus benefits from having a place to stay and a steady sugar supply from the plant, and the plant benefits because the fungus increases the surface area for water uptake and supplies the plant with certain minerals.

22. **(B) is correct.** Plants that have the staminate and carpelate flowers on the same individual plant are said to be monoecious. One example of a plant like this is the corn plant. Dioecious plants have the staminate flowers and carpelate flowers on different plants.

23. **(B) is correct.** The leaf is the main photosynthetic organ of the plant, and it varies quite a bit in form, from plant to plant, but most plant leaves consist of a blade (which is the leaf part of the leaf) and a petiole, which joins the leaf to a node on the stem.

24. **(B) is correct.** The phloem transports the food made in mature leaves to the roots and other nonphotosynthetic parts of the plant, like newly developing leaves and fruits. The xylem is responsible for carrying water and minerals up through the plant from the roots. Those are the two types of vascular tissue in plants.

25. **(E) is correct.** It has been proved that blue light is most effective in initiating germination in seeds, the opening of the stomata, and phototropism (when a plant bends toward or away from light).

Free-Response Question

1. (a) The elongation of the stem in plants is an important process in the plant life cycle because it enables the plant to reach its full size and complete development of the sporophyte stage. The sporophyte needs to complete develop-

ment in order to produce reproductive structures—flowers, which enable it to reproduce.

The apical meristems, which are located at the ends of a plant's terminal buds, are the sites of shoot growth. The apical meristems are masses of dividing cells that give rise to primary meristems; protoderm, procambrium, and ground meristem, which will later develop into the three vascular tissues of plants, vascular tissue, dermal tissue, and ground tissue. The apical meristem produces internodes (which contribute to plant height through both cell elongation and cell division) and leaf-bearing nodes. Both auxins and gibberellins are the hormones involved in stem elongation in plants. Auxins stimulate growth when they are present in low concentrations, and gibberellin concentration must be high.

(b) It is important for the plant to orient itself so that its roots reach into the earth and extract water and minerals, while its leaves point up to the sun so that they can trap the light energy from the sun (to use in photosynthesis) and so that transpiration can freely take place on the leaves. *Gravitropism* is the term used to describe plants growth in response to the force of gravity. Roots show positive gravitropism—they grow into the earth with gravity—whereas shoots show negative gravitropism—that is, they grow away from the force of gravity toward the sun. The major hormone involved in gravitropism is auxins. It has not yet been determined exactly how auxins influence the way roots and shoots grow, but it is theorized that certain dense molecules settle with gravity to one end of the plant root, and auxin accumulates as a result of this. The accumulation of auxin prevents cell elongation on that side of the root, and cells on the upper side elongate so that the root curves down, into the earth.

(c) Photoperiodism is defined as any physiological response to a photoperiod (meaning, a specific length of daylight, or night length). One example of a photoperiodism is flowering. It used to be thought that plant flowering depended on the length of the daylight, but then scientists realized that it is actually night length that determines when a plant will flower. They realized this by interrupting the day with a dark period and seeing that the plant still flowered. However, when they interrupted the night time with a brief period of light, the plant did not flower. Auxins are involved in photoperiodism, but it is not clear yet exactly what their role is.

This response uses the following key terms in context, showing the writer's knowledge of their meanings and relatedness:

sporophyte	*ground meristem*	*auxins*
flower	*vascular tissue*	*gibberellins*
apical meristem	*dermal tissue*	*gravitropism*
primary meristems	*ground tissue*	*photoperiodism*
protoderm	*internodes*	
procambrium	*leaf-bearing nodes*	

It also shows an understanding of the following processes: plant growth via stem elongation, gravitropism, and photoperiodism.

CHAPTER 7

Animal Form and Function

An Introduction to Animal Structure and Function: Functional Anatomy: An Overview

▌ **Anatomy** is defined as the study of the structure of an organism.

▌ **Physiology** is defined as the study of the functions of an organism.

▌ **Tissues** are groups of cells that have a common structure and function. The four types of tissue are listed below.

1. **Epithelial tissue** occurs in sheets of tightly packed cells, covers the body, lines the organs of the body, and acts as a protective barrier. One side of an epithelial cell is always bound to an underlying supportive surface called the basement membrane. The outside surface is facing either air or a fluid environment.

2. **Connective tissue** mainly supports and binds other tissues. It consists of scattered cells within an extracellular matrix. Some connective tissues are cartilage, tendons, ligaments, bone, and blood.

3. The functional unit of **nervous tissue** is the nerve cell, or neuron. This tissue senses stimuli and transmits signals from one part of the body to other neurons, glands, muscles, and the brain.

4. **Muscle tissue** is composed of long cells called muscle fibers. Muscle fibers contract when they are stimulated by a nerve impulse. This is the most abundant tissue in most animals. There are three types of muscle—skeletal muscle, cardiac muscle, and smooth muscle.

▌ **Organs** are organized groups of tissue. **Organ systems** are organs that work together cooperatively to do a common function.

Regulating the Internal Environment

▌ The **interstitial fluid** refers to the internal environment of an animal. **Homeostasis** is the state of internal balance in an animal.

▌ Homeostatic control systems have three components—a **receptor**, a **control center**, and an **effector**. The receptor detects a change; the control center processes information and directs the effector to make an appropriate response.

▌ In **negative feedback systems**, a change in the variable being monitored triggers a change in the control center that prevents further change in the variable or brings the variable back within desirable parameters.

Introduction to the Bioenergetics of Animals

▌ The **metabolic rate** of an animal is defined as the amount of energy it uses in a unit of time.

- **Endothermic** animals are warmed by the heat generated by their metabolism, and **ectothermic** animals do not produce enough heat by metabolism to influence their body temperature.
- The **basal metabolic rate** of an animal is defined as its metabolic rate when it is at rest, is experiencing no stress, and has an empty stomach.

Animal Nutrition: Nutritional Requirements

- An adequate diet supplies three things: fuel in the form of chemical energy, the organic raw materials for biosynthesis, and essential nutrients.
- The **essential nutrients** required by an animal are those that must be obtained in preassembled organic form because the animal cannot produce them.
- About half of the 20 **essential amino acids** must be obtained from food. There are also **essential fatty acids** which animals can not make and must ingest.
- **Vitamins** are organic molecules that are required in the diet in small amounts. They are used as co-factors in enzyme-controlled biochemical reactions.
- **Minerals** are simple inorganic nutrients that are also required in the diet in small amounts.

Food Types and Feeding Mechanisms

- **Herbivores** eat mainly autotrophs.
- **Carnivores** eat other animals.
- **Omnivores** eat both plants and animals.

Overview of Food Processing

- **Ingestion** is the act of taking in food, and it is the first stage in the processing of food.
- **Digestion** is the second stage of the processing of food. It is the breakdown of food into small molecules capable of being absorbed by the cells of the body.
- **Enzymatic hydrolysis** is the reaction by which macromolecules are broken up. It involves the addition of water.
- **Absorption** is the stage in food processing when the body's cells take up small molecules from the digestive tract.
- **Elimination** occurs when the undigested material passes out of the digestive tract.
- **Intracellular digestion** occurs within a cell enclosed by a protective membrane. Sponges digest their food this way.
- **Extracellular digestion** is carried out by most animals; in this type of digestion, food is broken down outside of cells.
- Many simple animals have a **gastrovascular cavity**, where digestion takes place. These simple animals have a single opening through which food enters and waste is eliminated.
- More complex animals have **complete digestive tracts** (**alimentary canals**), which are one-way digestive tubes that begin with the mouth at one end of terminate in the anus at the other.

The Mammalian Digestive System

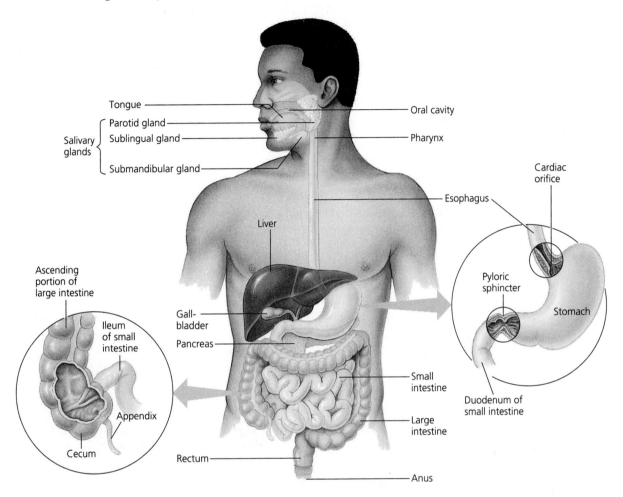

The Human Digestive System

▌ When food is in the mouth, or oral cavity, a nervous reflex occurs which causes **saliva** to be secreted into the mouth. Saliva lubricates the food and contains the enzyme **salivary amylase**, which hydrolyzes starch and glycogen into smaller polysaccharides and the disaccharide maltose.

▌ During chewing, food is shaped into a ball called a **bolus**. After being swallowed, the bolus enters the **pharynx**—a junction that opens to the esophagus and the trachea. During swallowing, the **epiglottis** (a flap made of cartilage) moves to cover the trachea. This will divert the food to go down the esophagus.

▌ The **esophagus** moves food from the pharynx down to the stomach through **peristalsis**—rhythmic waves of contraction by smooth muscle in the walls of the esophagus.

▌ The **stomach** is in the upper abdominal cavity, and its functions include storing food and secreting gastric juice. **Gastric juice** contains hydrochloric acid, which is very acidic (pH of about 2). Gastric juice breaks down the extracellular matrix of meat and plant materials, and it also kills most of the bacteria ingested with the food.

- **Pepsin** is an enzyme in gastric juice that begins to hydrolyze proteins into smaller polypeptides. Pepsin is secreted in an inactive form called **pepsinogen**, which is activated by hydrochloric acid in the stomach.
- The result of digestion in the stomach is a substance called **acid chyme**. The acid chyme is shunted from the end of the stomach into the beginning of the small intestine via the **pyloric sphincter**.
- The **small intestine** is the longest section of the alimentary canal. The beginning of the small intestine is the site of most of the hydrolysis of macromolecules, and the rest of the small intestine is responsible for the absorption of nutrients into the blood.
- The first section of the small intestine is known as the **duodenum**. In the duodenum, the acid chyme mixes with secretions from the pancreas (**bicarbonate**, which acts as a buffer against acid chyme), the liver (**bile**, which contains bile salts—detergents that aid in digestion), the gallbladder, and the intestinal wall itself.
- Here is how particular macromolecules are broken down in the small intestine.
    1. **Carbohydrates**—The breakdown of starch and glycogen begins with salivary amylase in the mouth. In the small intestine **pancreatic amylases** break starch, glycogen, and small polysaccharides into disaccharides. The breakdown of these disaccharides occurs at the wall of the intestinal epithelium, and then the monosaccharides are quickly absorbed.
    2. **Proteins**—Pepsin begins the breakdown of proteins in the stomach, and in the small intestine, **trypsin** and **chymotrypsin** break polypeptides into smaller chains. **Dipeptidases**, **carboxypeptidase**, and **aminopeptidase** break apart proteins into amino acids.
    3. **Nucleic acids**—The breakdown of nucleic acids is similar to that of proteins. In the small intestine, nucleases break them down into nucleosides, nitrogenous bases, sugars, and phosphate groups.
    4. **Fats**—Digestion of fats starts in the small intestine. Bile salts coat the fat droplets and keep them from coalescing (in **emulsification**), and **lipase** hydrolyzes them.
- Most absorption of nutrients occurs in the small intestine, and the epithelial lining of the small intestine has folds called **villi**, which in turn bear projections called **microvilli**—both of which radically increase the surface area available for absorption.
- In each villus is a set of tiny blood vessels called capillaries and a lymph vessel called a **lacteal.**
- Monosaccharides, such as glucose, cross via passive diffusion, whereas amino acids and dipeptides are pumped across in active transport.
- The lacteal will absorb small fatty acids.
- The capillaries and veins that drain the nutrients away from the villi all join the **hepatic portal vessel**, which brings them to the **liver**. The liver metabolizes the organic molecules in various ways.

- Some hormones involved in digestion are **gastrin**, which stimulates the secretion of gastric juice; and **enterogastrones**, such as **secretin** and **cholescystokinin** (**CCK**), that are secreted by the walls of the duodenum and that prompt the digestion of various macromolecules.
- The **large intestine**, also called the **colon**, is connected to the small intestine by a sphincter. The point of the connection is the site of the **cecum**, a small pouch with an extension called the **appendix**.
- The main function of the large intestine is to compact waste and recover water from it that can be returned to the body. The wastes become more solid as they travel along and form feces.
- At the end of the colon is the **rectum**, where feces are stored until they are eliminated.

Circulation and Gas Exchange: Circulation in Animals

- Both open and closed circulatory systems have **blood** (a circulatory fluid), **vessels** (tubes through which blood moves), and a **heart** (a structure that pumps the blood).
- In **open circulatory systems**, blood bathes the organs directly. The blood and lymph combined are called hemolymph, and a heart pumps hemolymph into cavities called sinuses.
- In **closed circulatory systems**, blood is contained within vessels and pumped around the body; the blood is separate from the interstitial fluid.
- Humans have a closed circulatory system called the **cardiovascular system**. The heart has **atria** (chambers that receive blood returning to the heart) and **ventricles** (chambers that pump blood out of the heart).
- The main types of blood vessels in humans are the **arteries, veins**, and **capillaries**. Arteries carry blood away from the heart and branch into smaller arterioles. Then capillaries network to form capillary beds. These capillary beds converge into venules, which converge into veins, which carry the blood back to the heart.

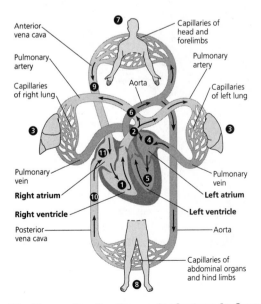

The Mammalian Cardiovascular System: An Overview

▌ Here are the steps of double circulation in mammals.
1. The right ventricle pumps blood to the lungs through the pulmonary arteries.
2. The blood flows through capillary beds in the lungs and picks up oxygen and releases CO_2.
3. The blood returns to the left atrium of the heart via pulmonary veins.
4. Then it continues to the left ventricle.
5. Then it leaves the heart via the aorta, which sends blood to arteries throughout the body.
6. The blood enters capillary beds in the neck, head, and arms.
7. The blood enters capillary beds in the abdomen and legs, giving up oxygen and picking up CO_2 from cell respiration.
8. The capillaries form venules, and blood from the neck, head, and arms travels back to veins and back to the right atrium via the anterior vena cava.
9. Blood from the legs and trunk travels through the posterior vena cava back to the right atrium.

▌ The complete cycle of contraction and relaxation of the heart is called the **cardiac cycle**. The contraction phase is called **systole**, and the relaxation phase is called **diastole**.

▌ **Heart rate** is the rate of contraction per minute, and the **stroke volume** is the amount of blood pumped by the left ventricle during each contraction.

▌ There are four heart valves and two **atrioventricular (AV) valves** between each atrium and ventricle, which prevent the backflow of blood into the atria; there are also two **semilunar valves**—one located at the entrance to the pulmonary artery and the second at the entrance to the aorta.

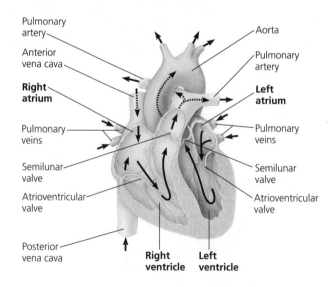

The Mammalian Heart

Regulation of the Heartbeat

- The **sinoatrial (SA) node** is the pacemaker of the heart. It is located in the upper wall of the right atrium. It sets the rate at which cardiac muscle cells contract.
- The **AV**, or **atrioventricular node**, located in the lower wall of the right atrium, delays the impulses from the SA node to allow the atria to completely empty before the ventricles contract.

The Lymphatic System and Blood

- The **lymphatic system** is responsible for returning lost fluid and proteins from the blood back into the blood. **Lymph** is the lost fluid and proteins carried in the lymphatic system. Along a lymph vessel are **lymph nodes** that filter lymph and attack viruses and bacteria, playing an important role in immunity.
- **Blood** is actually a connective tissue made up of many kinds of cells in a liquid matrix called plasma. **Plasma** is mostly water, but it also contains ions, electrolytes, and plasma proteins. It also carries nutrients, metabolic wastes, gases, and hormones. In addition, blood plasma carries:
    1. **red blood cells (erythrocytes)**, which transport oxygen via hemoglobin (an iron-containing protein);
    2. **white blood cells (leukocytes)**, which are part of the immune system;
    3. and **platelets**, which are fragments of cells responsible for blood clotting.
- Blood contains a soluble plasma protein called **fibrinogen**, which forms clots when it is converted to its active form, **fibrin**.

Gas Exchange in Animals

- **Gas exchange**, or **respiration**, is the uptake of molecular oxygen (O_2) from the environment, and the discharge of carbon dioxide to the environment. The part of the animal where gas exchange takes place is called the **respiratory surface**.
- **Gills** are respiratory organs in aquatic animals. Water flows through them, and blood flowing through capillaries within the wall of the gill picks up oxygen from the water. Blood flows in a direction opposite to the flow of water. This is called **countercurrent exchange**, and it maximizes the absorption of oxygen.
- Insects have **tracheal systems**, which are made up of air tubes that branch through the body and open to the outside. They extend to almost all cells, and gas exchange occurs directly across the epithelial membrane inside the tracheal walls.
- The **larynx** (voicebox) is the upper part of the respiratory tract. It is a tube with cartilage-reinforced walls that leads to the **trachea** (windpipe). The trachea divides into two **bronchi**, each of which leads to a lung. In the lungs the bronchi branch into **bronchioles**, and at their tips, the bronchioles end in clusters of air sacs called **alveoli**, the sites of gas exchange.
- **Breathing** is the inhalation and exhalation of air that ventilates lungs. Breathing involves movement of the **diaphragm**—a dome-shaped muscle separating

the thoracic cavity from the abdominal cavity. Lung volume increases when the rib muscles and diaphragm contract.

▌ The diffusion of a gas depends on **partial pressure**. Gases always diffuse from regions of higher partial pressure to regions of lower partial pressure.

▌ **Hemoglobin** is the respiratory pigment found in almost all vertebrates. It consists of four subunits, each of which is a heme group with an embedded iron atom. The iron atom binds O_2, so each hemoglobin can carry 4 oxygen molecules.

▌ A lowering of the pH in blood lowers the affinity of hemoglobin for oxygen, and oxygen dissociates. This is called the **Bohr shift**.

▌ CO_2 is most commonly carried in the blood in the form of bicarbonate ion. Less commonly it is transported via hemoglobin and in solution in the blood plasma.

The Body's Defenses: Nonspecific Defenses Against Infection

▌ Skin and the mucous membranes cover the surface and line the openings of the animal body, and they provide an external barrier against infecting agents.

▌ Microbes that get through the skin—for instance, in a cut—encounter certain types of white blood cells called **neutrophils** that ingest and destroy them in a process called **phagocytosis**.

▌ **Monocytes** are another type of phagocytotic leukocyte. They migrate into tissues and develop into macrophages, which are giant phagocytotic cells.

▌ **Eosinophils** are leukocytes that defend against parasitic invaders such as worms by positioning themselves near to the parasites wall and discharging hydrolytic enzymes.

▌ The **inflammatory response** occurs when physical injury occurs to a tissue, and occurs in response to chemical signals. For example, histamines are released by basophils and mast cells (two types of leukocytes) in response to injury. Histamines trigger the dilation and permeability of nearby capillaries. This aids in delivering clotting agents to the injured area.

How Specific Immunity Arises

▌ Vertebrates have two types of lymphocytes: **B lymphocytes** (**B cells**), which proliferate in the bone barrow, and **T lymphocytes** (**T cells**), where lymphocytes mature in the thymus. They circulate through the blood and lymph, and both recognize particular microbes and are said to show specificity.

▌ **Antigens** are foreign molecules that elicit a response by lymphocytes.

▌ **Antibodies** are proteins secreted by B cells in an immune response.

▌ **Antigen receptors** are located on the antigen and allow B and T cells to recognize them. Antigen receptors on T cells are called **T cell receptors**, and they recognize antibodies specifically.

▌ When an antigen binds to a B or T cell, the lymphocyte becomes activated, and forms two clones of cells. One is made up of **effector cells**, which combat the antigen, and the other consists of **memory cells**, which are long-lived and bear receptors for the same antigen. This process is called **clonal selection**.

- When the body is first exposed to an antigen and a lymphocyte is activated, this is referred to as the **primary immune response**. The **secondary immune response** is faster and of greater magnitude.
- Lymphocytes and all other blood cells arise from stem cells in the bone marrow or the liver of a developing fetus.

Regulating the Internal Environment: An Overview of Homeostasis

- **Homeostasis** refers to an animal's ability to regulate its internal environment. **Thermoregulation** refers to how animals maintain their internal temperature, and **osmoregulation** refers to how they regulate solute balance and water content.
- **Conduction** is the transfer of heat between molecules of objects that are in direct contact with one another—for example, when an animal sits in water that is cooler than its body temperature.
- **Convection** is the transfer of heat through the movement of air or a liquid past a surface—for example, when a breeze causes heat loss from the surface of an animal.
- **Radiation** is the emission of electromagnetic waves by all objects that are warmer than absolute zero.
- **Evaporation** is the removal of heat from the surface of a liquid, as molecules leave the surface as gas.
- Thermoregulation takes place through the following processes.
  1. The adjustment of the rate of heat exchange between the animal and its environment—through insulating hair, feathers, and fat—is accomplished through **vasodilation** (an increase in diameter of blood vessels at the skin, which cools the blood) or **vasoconstriction** (the opposite of vasodilation);
  2. Evaporation across the skin (through panting or sweating);
  3. Behavioral responses (changes in location or posture);
  4. Alteration of the rate of metabolic heat production (only in endotherms).
- Mammals and birds are **endothermic**, whereas amphibians and reptile are **ectothermic**. Fishes and aquatic invertebrates are conformers—that is, they live in relatively stable environments and can accommodate some slight change in body temperature if the environment is altered.

Water Balance and Waste Disposal

- Most metabolic wastes must be excreted from the body. One of the most important waste products are **nitrogen-containing breakdown products** of proteins and nucleic acids.
- Enzymes remove nitrogen from these compounds to create **ammonia**. Some animals excrete ammonia directly into water, where it becomes diluted. Others convert it first to **urea** in the liver, where ammonia is combined with carbon dioxide in an endergonic reaction, or to **uric acid**. Uric acid is more energetically expensive to produce, but it is insoluble in water and can be excreted as a paste or crystals.

- Most excretory systems produce urine in a two-step process. First the body fluid (blood or hemolymph) is collected, and then the composition of the fluid is adjusted by **selective reabsorption of** solutes.

- Insects and terrestrial arthropods like the grasshopper have **Malpighian tubules** that remove nitrogenous wastes. They open into the digestive tract and dead-end at points in the hemolymph. The tubules secrete nitrogenous wastes and salts into the lumen, and water follows by osmosis.

- Mammals have two **kidneys**, and each is supplied with a **renal artery** and a **renal vein**. **Urine** leaves the kidneys through the **ureters**, which drain into the urinary bladder. Urine is expelled from the body through the **urethra**.

- The kidney has two regions, the **outer renal cortex** and the **inner renal medulla**. These two regions are packed with nephrons, which are the functional units of the kidney.

- **Nephrons** are made up of a single long tubule and the **glomerulus**, a ball of capillaries. At the end of the tubule is the **Bowman's capsule**, a c-shaped capsule that surrounds the glomerulus.

- The filtrate flows through the **proximal tubule**, the descending loop of Henle, the **loop of Henle**, the ascending loop of Henle, and the **distal tubule**. The distal tubule empties into a **collecting duct**, which receives wastes from many nephrons. The filtrate empties into the renal pelvis.

- In the human kidney, most of the nephrons are **cortical nephrons**; these are in the renal cortex. The rest are **juxtamedullary nephrons**, with long loops of Henle that extend into the renal medulla.

- Capillaries called **afferent arterioles** are associated with the nephrons, and as they leave the glomerulus, the capillaries converge into an **efferent arteriole**. This vessel subdivides again to form **peritubular capillaries**, which surround the proximal and distal tubules.

There are five main steps in the **transformation of blood filtrate to urine**, as shown below.

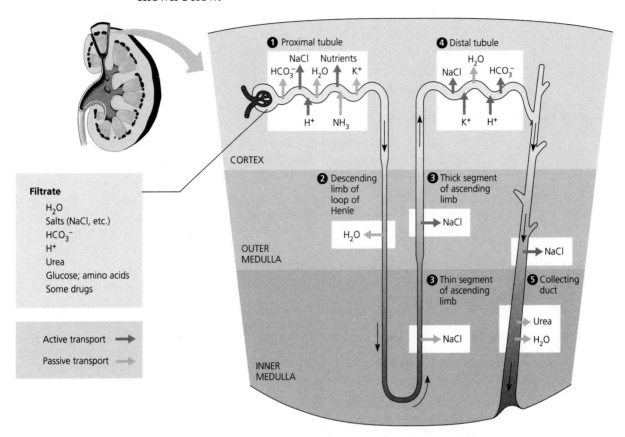

The Nephron and Collecting Duct: Regional Functions of the Transport Epithelium

1. In the proximal tubule, secretion and reabsorption changes the volume and composition of the filtrate. The pH of body fluids is controlled, and bicarbonate is absorbed, as are NaCl and water.
2. In the descending loop of Henle, reabsorption of water continues.
3. In the ascending loop of Henle, the filtrate loses salt without giving up water and becomes more dilute.
4. In the distal tubule, K⁺ and NaCl levels are regulated, as is filtrate pH.
5. The collecting duct carries the filtrate through the medulla to the renal pelvis, and the filtrate becomes more concentrated by the movement of salt.

Antidiuretic hormone is an important hormone in the regulation of water balance. It is produced in the hypothalamus and stored in and released from the pituitary gland. Two other hormones involved in regulation of water balance are **angiotensin** and **aldosterone**.

Chemical Signals in Animals

Hormones are chemical signals released into body fluids that communicate messages around the body.

Target cells are those cells equipped to respond to hormones.

An Introduction to Regulatory Systems

▌ The **endocrine system** of an animal is the sum of all its hormone-secreting cells and tissues. Hormone-secreting organs are called endocrine glands.

▌ Many endocrine glands contain **neurosecretory cells**, which secrete hormones. Many chemicals act as both hormones and nervous system signals.

▌ **Feedback** is one important way by which the endocrine and nervous system are regulated.

Chemical Signals and Their Modes of Action

Chemical signals may bind to receptors on the plasma membranes of certain cells, and this triggers a **signal-transduction pathway**. A signal-transduction pathway consists of a series of molecular events that initiates a response to the signal. Alternately, the signal enters the target cell and binds to a receptor in the cell. The receptor then acts as a transcription factor, causing a change in gene expression.

The Vertebrate Endocrine System

▌ Hormones in the body can affect one tissue, a few tissues, or most of the tissues in the body (like the sex hormones), or they may affect other endocrine glands (these last are referred to as **tropic hormones**).

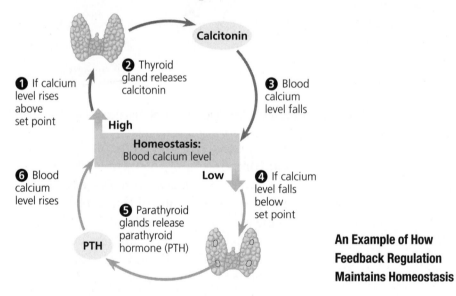

An Example of How Feedback Regulation Maintains Homeostasis

Animal Reproduction: Overview of Animal Reproduction

▌ **Sexual reproduction** is the creation of offspring by the fusion of haploid gametes to form a zygote. The female gamete is the ovum, and the male gamete is the spermatozoon.

▌ **Asexual reproduction** is reproduction in which all genes come from one parent; there is no fusion of egg and sperm. Asexual reproduction can occur by **binary fission**, which is the separation of a parent into two or more individuals of about the same size.

▌ **Budding** is another form of asexual reproduction, in which new individuals bud off of the parent. Budding occurs in yeast cells.

- **Fragmentation** is a form of asexual reproduction in which an individual breaks into several pieces, all of which then form complete adults. Regeneration, the re-growth of body parts, is a necessary part of fragmentation.
- **Parthenogenesis** is the process in which an egg develops without being fertilized.
- **Hermaphroditism** exists when each individual has both male and female reproductive systems.

Mechanisms of Sexual Reproduction

- **Fertilization** is the union of sperm and egg. **External fertilization** occurs when eggs are shed by the female and fertilized by the male outside the female's body, and **internal fertilization** occurs when sperm are deposited in the female reproductive tract and fertilization occurs within the tract.
- **Gonads** are the organs that produce gametes in most animals.

Mammalian Reproduction

- The male's external reproductive organs are the **scrotum** and **penis**, and the internal organs are gonads (which produce gametes and hormones), accessory glands (which secrete necessary fluids), and ducts (which carry sperm and glandular secretions).
- The testes are made up of many highly coiled tubules surrounded by connective tissue. The tubules are **seminiferous tubules**, the sites of sperm production. In between the tubules are **Leydig cells**, which produce testosterone and other androgens.
- The testes are held in the **scrotum**, which is located outside the lower abdominal pelvic cavity.
- The sperm passes from the seminiferous tubules into the **epididymis**. During ejaculation, the sperm is propelled through the **vas deferens**, which ultimately meets up with a duct from the **seminal vesicle** and forms an ejaculatory duct, which opens into the urethra.
- The seminal vesicles, the **prostate gland**, and the **bulbourethral gland** all contribute secretions to create semen. These secretions supply necessary nutrients and a medium for the sperm cells.
- The **penis** is composed of three masses of spongy tissue derived from modified veins and capillaries.
- The female gonads are the two **ovaries**. Each ovary contains many microscopic follicles.
- **Follicles** consist of one egg surrounded by one or more layers of follicle cells, which help to develop, nourish, and protect the egg cell. All of the follicles a woman has are formed before her birth. One follicle matures and releases its egg cell during each menstrual cycle.
- The follicle cells also produce **estrogens**, the female hormones.
- The egg cell is released from the follicle during ovulation. The remaining follicle tissue heals and grows in the ovary to form a body called a **corpus luteum**. This secretes estrogens and progesterone. Progesterone helps to maintain the uterine wall during pregnancy. If the egg cell isn't fertilized, the corpus luteum disintegrates.

- The egg cell is released into the **oviduct**, or **fallopian tube**, and cilia lining the fallopian tube convey the egg cell down to the **uterus**. The inner lining of the uterus is called the **endometrium**.
- At the base of the uterus is the **cervix**, which leads to the **vagina**, the canal through which a baby is born.
- **Spermatogenesis** is the production of mature sperm cells, and it occurs in the seminiferous tubules. The cells that give rise to sperm are called **spermatogonia**. They undergo meiosis and differentiation eventually to form mature, motile sperm.
- The head of sperm cells is tipped with an **acrosome**, which secretes enzymes that help the sperm penetrate the egg.
- **Oogenesis** is the development of mature ova. **Oogonia** are the cells that develop into ova; they multiply and begin meiosis, but they stop at prophase I of meiosis I. These egg cells are called **primary oocytes**, which are quiescent until puberty. From puberty onward, FSH periodically stimulates a follicle to grow and its egg cell to complete meiosis I and begin meiosis II. This forms the **secondary oocyte**.
- Humans and other primates have **menstrual cycles**, and other mammals have **estrous cycles. Menstruation** occurs when the endometrium is shed from the uterus through the cervix and vagina.

  - The **menstrual flow phase** of the female cycle is the phase during which menstrual bleeding occurs.
  - The **proliferative phase** of the menstrual cycle is that during which the endometrium begins to regenerate and thicken.
  - In the **secretory phase**, the endometrium continues to thicken, and if an embryo has not implanted in the lining by the end of this phase, menstrual flow occurs.
  - The **ovarian cycle** parallels the menstrual flow cycle, and begins with the follicular phase, in which several follicles begin to grow.
  - At the end of the follicular phase, ovulation occurs, during which the secondary oocyte is released from the ovary.
  - During the **luteal phase** of the ovarian cycle, endocrine walls in the corpus luteum secrete hormones.

- **Pregnancy** (**gestation**) is the condition of carrying at least one embryo.

Animal Development: The Stages of Early Embryonic Development

- After fertilization, there are three successive stages in early development.

  1. **Cleavage**, which is a period of rapid mitotic cell division, partitions the cytoplasm of the zygote into smaller cells called **blastomeres**, each of which has its own nucleus. Continued cleavage leads to a ball of cells called a **morula**, and then a fluid-filled central cavity called the **blastocoel** forms within the morula to produce a **blastula**.
  2. **Gastrulation** is a drastic rearrangement of the cells in the blastula. In gastrulation, three (germ) cell layers are produced—the ectoderm (later forms the nervous system and outer layer of skin), the endoderm (later

develops to line the digestive tract), and the mesoderm (later develops into most organs and tissues).

▌ **Organogenesis** is the development of the three germ layers into the rudiments of organs.

Nervous Systems: An Overview of Nervous Systems

▌ **Sensory receptors** collect information about the world outside the body as well as processes inside the body.

▌ The **central nervous system** (CNS) consists of the brain and spinal cord, and the **peripheral nervous system** consists of the nerves that communicate motor and sensory signals throughout the rest of the body.

▌ **Motor output** is the conduction of signals from the CNS to **effector cells**, which are muscle or gland cells that carry out responses.

▌ The **neuron** is the functional unit of the nervous system. It is composed of a cell body which contains the nucleus and organelles; **dendrites**, which are cell extensions that receive incoming messages from other cells; and **axons**, which convey messages to other cells.

▌ Many axons are covered by an insulating fatty **myelin sheath**. At the end of the axons there are **synaptic terminals**, which relay signals from one neuron to another cell through chemical messengers called **neurotransmitters**.

▌ In nerve transmission, the transmitting cell is the **presynaptic cell**, and the receiving cell is the **postsynaptic cell**.

▌ A **simple nerve circuit** is the reflex arc, in which a sensory nerve receives information and passes it on to the spinal cored and then to a motor neuron, which signals an effector cell.

▌ **Ganglia** are clusters of nerve cells.

▌ **Glial cells** are supporting nerve cells, and they outnumber nerve cells in the body. Three important kinds of glial cells are **astrocytes,** which provide support for neurons; **oligodendrocytes,** which form myelin sheathes in the CNS; and **Schwann cells**, which form myelin sheaths in the peripheral nervous system (PNS).

The Nature of Nerve Signals

▌ **Membrane potential** describes the difference in electrical charge across a cell membrane.

▌ The membrane potential of a nerve cell at rest is called its **resting potential**. It exists because of differences in the ionic composition of the extracellular and intracellular fluids of the axonal membrane.

▌ Changes in the axonal membrane potential of a neuron are what give rise to **nerve impulses**. A stimulus first affects the membrane's permeability to ions, and this is a graded potential with a magnitude proportional to the size of the stimulus.

▌ An **action potential** (nerve impulse) is an all-or-none depolarization of the membrane of the nerve cell. It opens voltage-gated sodium channels, and Na^+ ions enter the cell, bringing the membrane potential to a positive value. The Na^+ gates then close, and the cell goes back to resting potential.

- Action potentials are propagated along the axon; **saltatory conduction**, which is the jumping of the nerve impulse between nodes of Ranvier (areas on the axon not covered by myelin sheath), which speeds up the conduction of the nerve impulse.

- The signal is conducted from the axon of a presynaptic cell to the dendrite of a postsynaptic cell via an **electrical** or **chemical synapse**.

- Electrical synapses occur via **gap junctions**.

- In chemical synapses, neurotransmitters are released by the presynaptic membrane into the synaptic cleft. They bind to receptors on the postsynaptic membrane and are then broken down by enzymes, or taken back up into surrounding cells.

- **Excitatory postsynaptic potential** (**EPSP**) is the electrical charge caused by the binding of the neurotransmitter to its receptor on the postsynaptic membrane.

- **Inhibitory postsynaptic potential** (**IPSP**) is the voltage charge associated with chemical signaling at an inhibitory synapse.

- **Acetylcholine** is a very common neurotransmitter, it can be inhibitory or excitatory. Other **neurotransmitters** are epinephrine and norepinephrine, dopamine, and serotonin.

Vertebrate Nervous Systems

- The **autonomic nervous system** transmits signals that regulate the internal environment by controlling smooth and cardiac muscle, including those in the gastrointestinal, cardiovascular, excretory and endocrine systems.

- The autonomic nervous system is composed of the **sympathetic division**—which when activated causes the heart to beat faster and adrenaline to be secreted (with all its effects)—and the **parasympathetic division**, which has the opposite effect when activated.

Sensory and Motor Mechanisms: Introduction to Sensory Reception

- **Mechanoreceptors** are receptors stimulated by physical stimuli, such as pressure, touch, stretch, motion, or sound.

- **Thermoreceptors** respond to either heat or cold and help maintain body temperature by keeping the core temperature stable.

- **Chemoreceptors** transmit information about solute concentration in a solution. Gustatory (taste) receptors and olfactory (smell) receptors are two types of chemoreceptors.

Photoreceptors and Vision

- **Compound eyes** (in insects and crustaceans) consist of up to several thousand light detectors called ommatidia, each of which has its own lens.

- **Single-lens eyes** are found in vertebrates and some invertebrates.

- The **eyeball** in single-lens eyes is made up of two outer layers, the **sclera** (which in the front of the eye is the **cornea**—responsible for acting as a lens) and the **choroid.** The eyeball also contains the **pupil**, which is the hole in the center of the **iris,** and the **retina**, which contains the photoreceptor cells.

- **Aqueous humor** fills the anterior cavity of the eye, and the **vitreous humor** fills the posterior cavity of the eye.
- The retina contains **rod cells** and **cone cells**, two types of photoreceptors.
- **Rhodopsin** is the light-absorbing pigment that triggers a signal-transduction pathway that ultimately leads to sight.

Hearing and Equilibrium

- There are three regions in the mammalian ear.
    1. The **outer ear**, which is the external **pinna** and **auditory canal.** These collect sounds and bring them to the **tympanic membrane** (eardrum), which separates the outer ear from the middle ear.
    2. The **middle ear**, in which vibrations are conducted through three small bones collectively called **ossicles** (individually, the **malleus, incus,** and **stapes**) and through the **oval window.**
    3. Then the vibrations are conducted to the **inner ear**, which consists of a labyrinth of channels lined by membrane and containing fluid, all situated in bone.

- The inner ear contains the **cochlea**, a two-chambered organ, which is involved in hearing.
- The **organ of Corti**, which is in the cochlea, contains the receptors of the ear, which are hair cells with hairs that project into the **cochlear duct.**
- The cochlea transduces the energy of the vibrating fluid into action potentials, in a wave that dissipates at the **round window.**
- Some organs in the inner ear are responsible for detecting body position and balance. These are the **semicircular canals**.

Chemoreception—Taste and Smell

Taste buds are modified epithelial cells situated on different parts of the tongue and mouth.

Movement and Locomotion

- **Locomotion** is the movement from place to place.
- **Hydrostatic skeletons** consist of fluid held under pressure in a closed body compartment.
- **Exoskeletons** are hard encasements on the surface of an animal, such as is found in the grasshopper. **Endoskeletons** consist of hard supporting elements buried within the soft tissues of an animal. An example is the human bony skeleton.
- **Skeletal muscle** is attached to bones and responsible for the movement of bones. It consists of long fibers, each of which is a single muscle cell. Each muscle fiber is a bundle of **myofibrils**, which in turn are composed of two kinds of **myofilaments; thin filaments** and **thick filaments.**
- **Skeletal muscle** is striated, and the basic contractile unit of the muscle is the **sarcomere.** The **Z lines** make up the border of sarcomeres; the **I band** is the

area near the end of the **sarcomere** where only thin filament exists; and the **A band** is the entire length of the thin filaments.

▎ During **muscle contraction**, the length of the sarcomere is reduced.

▎ The **sliding filament model** states that the thick and thin filaments slide past each other so that their degree of overlap increases. This is dependent on the interaction between the **actin** and **myosin** molecules that make up the thin and thick filaments.

▎ Muscle cells contract when stimulated by a **motor neuron**.

▎ To stimulate muscle contraction, an action potential in a motor neuron that makes a synaptic connection with the muscle cell releases acetylcholine at the **neuromuscular junction**. This depolarizes the muscle cell and triggers an action potential.

▎ The action potential spreads along **T tubules** (**transverse tubules**). This changes the permeability of the **sarcoplasmic reticulum** to calcium ions, and the newly released calcium ions bind to **troponin** and cause it to move, exposing the myosin sites on actin; the muscle contracts.

▎ **Fast-twitch muscle fibers** are used for fast, powerful contractions. **Slow-twitch muscle fibers** are used for slow, long-lasting contractions.

For Additional Review:

Consider how the immune system, the digestive system, the nervous system, the circulatory and respiratory systems, and the senses all contribute to homeostasis in animals. In doing so, connect the stimuli that engage these systems with the way systems respond to those stimuli.

Multiple-Choice Questions

1. Which of the following is required in all living things in order for gas exchange to occur?
   (A) Lungs
   (B) Gills
   (C) Moist membranes
   (D) Blood
   (E) Lymph

2. In animals, all of the following are associated with embryonic development EXCEPT
   (A) gastrulation
   (B) cleavage
   (C) depolarization
   (D) organogenesis
   (E) cell migration

3. Which of the following is most likely to result in a release of epinephrine (adrenaline) from the adrenal glands?
   (A) Falling asleep in front of the TV
   (B) Watching a golf tournament
   (C) Doing yoga
   (D) Taking a test without having studied for it
   (E) Being in the kitchen while dinner is being cooked

4. Oxygen is transported in human blood in which type of cells?
   (A) Erythrocytes
   (B) Leukocytes
   (C) Phagocytes
   (D) B cells
   (E) T cells

5. The proximal tubules in the kidney reabsorb most of which of the following compounds?
   (A) CO_2
   (B) O_2
   (C) H_2O
   (D) HCO_3^-
   (E) $C_6H_{12}O_6$

6. Salivary amylase, an enzyme secreted in saliva, begins the breakdown of which of the following substances?
   (A) Starch
   (B) Protein
   (C) Lipids
   (D) Nucleic acids
   (E) Polypeptides

Directions: The group of questions below consists of five lettered headings followed by a list of numbered phrases or sentences. For each numbered phrase or sentence select the one heading that is most closely related to it and fill in the corresponding oval on the answer sheet. Each heading may be used once, more than once, or not at all.

Questions 7–11
   (A) Ovary
   (B) Thyroid gland
   (C) Posterior pituitary gland
   (D) Adrenal medulla
   (E) Pineal gland

7. Releases hormones that raise blood glucose level, increase metabolic activities, and constrict blood vessels

8. Releases hormones that are involved in biological rhythms

9. Releases hormones that stimulate the contraction of the uterus and mammary gland cells

10. Releases hormones that stimulate growth of the uterine lining and promote the development of female secondary sex characteristics

11. Releases hormones that stimulate and maintain metabolic processes

12. Blood constitutes which of the following tissue types?
    (A) Epithelial tissue
    (B) Connective tissue
    (C) Nervous tissue
    (D) Vascular tissue
    (E) Glandular tissue

13. The three types of muscle in the body are
    (A) skeletal, cardiac, and smooth
    (B) skeletal, vascular, and smooth
    (C) skeletal, cardiac, and rough
    (D) cardiac, smooth, and rough
    (E) cardiac, smooth, and vascular

14. Which of the following is an example of a negative feedback system?
    (A) The movement of sodium across a membrane though an antiport, and the movement of potassium in the opposite direction through the same antiport
    (B) The pressure of the baby's head against the uterine wall during childbirth stimulates uterine contractions, which causes greater pressure against the uterine wall, which produces still more contractions.
    (C) The growth of a population of bacteria in a petri dish until it has used all of its nutrients, and its subsequent decline
    (D) A heating system in which the heat is turned off when the temperature exceeds a certain point, and is turned on when the temperature falls below a certain point
    (E) The progress of a chemical reaction until equilibrium is reached, and then the cycling back and forth of reactant to product

15. All of the following are fat-soluble vitamins EXCEPT
    (A) vitamin A
    (B) vitamin B
    (C) vitamin D
    (D) vitamin E
    (E) vitamin K

16. Which of the following has a diet that consists solely of autotrophs?
    (A) Omnivore
    (B) Carnivore
    (C) Herbivore
    (D) Trendavore
    (E) Supravore

17. The four stages of food processing are ingestion, digestion, absorption, and
    (A) incorporation
    (B) circulation
    (C) elimination
    (D) filtration
    (E) cellular uptake

18. Hydras possess which of the following types of digestive system?
    (A) Food vacuole
    (B) Complete digestive tract
    (C) Alimentary canal
    (D) Lumen
    (E) Gastrovascular cavity

19. Which of the following is the site of the production of bile?
    (A) Gall bladder
    (B) Small intestine
    (C) Prostate
    (D) Pancreas
    (E) Liver

20. The primary sites of carbohydrate digestion are which of the following structures?
    (A) Mouth and large intestine
    (B) Mouth and stomach
    (C) Stomach and small intestine
    (D) Mouth and small intestine
    (E) Small intestine and colon

21. Pepsin in the stomach is primarily responsible for the breakdown of which type of molecule?
    (A) Starch
    (B) Protein
    (C) Lipids
    (D) Nucleic acids
    (E) Glycogen

22. Which of the following structures is primarily responsible for reabsorbing water from the lumen?
    (A) Small intestine
    (B) Nephron
    (C) Glomerulus
    (D) Colon
    (E) Cecum

23. Insects and other arthropods have which of the following circulatory fluids?
    (A) Lymph
    (B) Hemoglobin
    (C) Blood
    (D) Hemolymph
    (E) Heterolymph

24. A body plan in which blood bathes the organs directly is termed
    (A) an open circulatory system
    (B) a closed circulatory system
    (C) a cardiovascular system
    (D) a gastrovascular system
    (E) a gastrovascular cavity system

25. Which of the following carry blood away from the heart?
    (A) Venules
    (B) Veins
    (C) Arteries
    (D) Capillaries
    (E) Atria

26. In the mammalian heart, the sinoatrial node (SA node) is responsible for which of the following functions?
    (A) Delaying the nerve impulse to the walls of the ventricle
    (B) Controlling the atrioventricular valve

(C) Controlling the semilunar valves
(D) Setting the rate and timing of cardiac muscle contraction
(E) Monitoring stroke volume

27. Fluid and proteins lost from the capillaries are returned to the blood via
(A) the venuous system
(B) the arteriole system
(C) the lymphatic system
(D) capillary beds
(E) the hemolymph system

28. All of the following are components of blood EXCEPT
(A) red blood cells
(B) white blood cells
(C) platelets
(D) leukocytes
(E) free amino acids

29. Red blood cells are derived from which of the following tissues?
(A) The heart
(B) The blood vessels
(C) Bone
(D) Muscles
(E) Masses of other blood cells

30. Grasshoppers exhibit which of the following types of respiratory systems?
(A) Lungs
(B) Countercurrent system
(C) Tracheal system
(D) Malpighian system
(E) Vessel system

31. In the blood, carbon dioxide is primarily transported in what way?
(A) By hemoglobin
(B) By hemocyanin
(C) As carbon monoxide
(D) As bicarbonate
(E) In erythrocytes

32. All of the following are first-line barriers against infectious agents EXCEPT
(A) skin
(B) nasal membranes
(C) saliva
(D) mucous secretions
(E) phagocytes

33. An immune response to a specific antigen generates the production of which type of cell that launches an attack the next time that same antigen infects the body?
(A) Effector cells
(B) Memory cells
(C) T cells
(D) B cells
(E) Antibodies

34. All of the following are ways by which organisms exchange heat EXCEPT
(A) transference
(B) conduction
(C) convection
(D) radiation
(E) evaporation

35. Which of the following animals is an ectoderm?
(A) Human being
(B) Snake
(C) Bird
(D) Monkey
(E) Coatimundi

36. The Malpighian tubules are the organs that constitute the excretory system of which of the following animals?
(A) Birds
(B) Human beings
(C) Fish
(D) Insects
(E) Hydra

37. The ball of capillaries that is associated with the nephron and associated with filtration in the kidney is
    (A) the Bowman's capsule
    (B) the loop of Henle
    (C) the proximal tubule
    (D) the glomerulus
    (E) the distal tubule

Directions: The group of questions below consists of five lettered headings followed by a list of numbered phrases or sentences. For each numbered phrase or sentence select the one heading that is most closely related to it and fill in the corresponding oval on the answer sheet. Each heading may be used once, more than once, or not at all.

Questions 38–42
    (A) Vitreous humor
    (B) Cone cells
    (C) Eustachian tube
    (D) Cochlea
    (E) Taste buds

38. A photoreceptor

39. A coiled organ that is involved in hearing

40. Constitutes most of the volume of the eye

41. Equalizes the pressure between the middle ear and the atmosphere

42. Receptors that can be stimulated by a broad range of chemicals

43. Muscle cell contraction occurs via
    (A) contraction of the A band
    (B) contraction of the I band
    (C) contraction of the Z lines
    (D) the sliding of the thin filaments by the thick filaments
    (E) the contraction of the sarcoplasmic reticulum

44. The succession of rapid cell division that follows fertilization is called
    (A) gastrulation
    (B) cleavage

    (C) morulation
    (D) involution
    (E) polarization

45. The circuit of a sensory neuron, the spinal cord, a motor neuron, and an effector cell constitutes a
    (A) presynaptic sequence
    (B) reflex arc
    (C) nerve circuit
    (D) nerve impulse
    (E) saltatory conduction system

46. Which of the following is released into the synaptic cleft and acts as an intracellular messenger?
    (A) Sodium
    (B) Chloride
    (C) Neurotransmitter
    (D) Action potential
    (E) Voltage gradient

47. An egg cell surrounded by one or two layers of cells is called a
    (A) follicle
    (B) corpus luteum
    (C) oviduct
    (D) endometrium
    (E) uterus

The Princeton Review.

Completely darken bubbles with a No. 2 pencil. If you make a mistake, be sure to erase mark completely. Erase all stray marks.

1.
YOUR NAME:
(Print)

Last   First   M.I.

SIGNATURE: _____ **DATE:** ___/___/___

HOME ADDRESS:
(Print)

Number and Street

City   State   Zip Code

PHONE NO.: _____

IMPORTANT: Please fill in these boxes exactly as shown on the back cover of your test book.

2. TEST FORM

3. TEST CODE

4. REGISTRATION NUMBER

5. YOUR NAME

First 4 letters of last name

FIRST INIT   MID INIT

6. DATE OF BIRTH

Month	Day	Year
JAN		
FEB		
MAR		
APR		
MAY		
JUN		
JUL		
AUG		
SEP		
OCT		
NOV		
DEC		

7. GENDER
○ MALE
○ FEMALE

The Princeton Review.

1. Ⓐ Ⓑ Ⓒ Ⓓ
2. Ⓐ Ⓑ Ⓒ Ⓓ
3. Ⓐ Ⓑ Ⓒ Ⓓ
4. Ⓐ Ⓑ Ⓒ Ⓓ
5. Ⓐ Ⓑ Ⓒ Ⓓ
6. Ⓐ Ⓑ Ⓒ Ⓓ
7. Ⓐ Ⓑ Ⓒ Ⓓ
8. Ⓐ Ⓑ Ⓒ Ⓓ
9. Ⓐ Ⓑ Ⓒ Ⓓ
10. Ⓐ Ⓑ Ⓒ Ⓓ
11. Ⓐ Ⓑ Ⓒ Ⓓ
12. Ⓐ Ⓑ Ⓒ Ⓓ
13. Ⓐ Ⓑ Ⓒ Ⓓ
14. Ⓐ Ⓑ Ⓒ Ⓓ
15. Ⓐ Ⓑ Ⓒ Ⓓ
16. Ⓐ Ⓑ Ⓒ Ⓓ
17. Ⓐ Ⓑ Ⓒ Ⓓ
18. Ⓐ Ⓑ Ⓒ Ⓓ
19. Ⓐ Ⓑ Ⓒ Ⓓ
20. Ⓐ Ⓑ Ⓒ Ⓓ

21. Ⓐ Ⓑ Ⓒ Ⓓ
22. Ⓐ Ⓑ Ⓒ Ⓓ
23. Ⓐ Ⓑ Ⓒ Ⓓ
24. Ⓐ Ⓑ Ⓒ Ⓓ
25. Ⓐ Ⓑ Ⓒ Ⓓ
26. Ⓐ Ⓑ Ⓒ Ⓓ
27. Ⓐ Ⓑ Ⓒ Ⓓ
28. Ⓐ Ⓑ Ⓒ Ⓓ
29. Ⓐ Ⓑ Ⓒ Ⓓ
30. Ⓐ Ⓑ Ⓒ Ⓓ
31. Ⓐ Ⓑ Ⓒ Ⓓ
32. Ⓐ Ⓑ Ⓒ Ⓓ
33. Ⓐ Ⓑ Ⓒ Ⓓ
34. Ⓐ Ⓑ Ⓒ Ⓓ
35. Ⓐ Ⓑ Ⓒ Ⓓ
36. Ⓐ Ⓑ Ⓒ Ⓓ
37. Ⓐ Ⓑ Ⓒ Ⓓ
38. Ⓐ Ⓑ Ⓒ Ⓓ
39. Ⓐ Ⓑ Ⓒ Ⓓ
40. Ⓐ Ⓑ Ⓒ Ⓓ

41. Ⓐ Ⓑ Ⓒ Ⓓ
42. Ⓐ Ⓑ Ⓒ Ⓓ
43. Ⓐ Ⓑ Ⓒ Ⓓ
44. Ⓐ Ⓑ Ⓒ Ⓓ
45. Ⓐ Ⓑ Ⓒ Ⓓ
46. Ⓐ Ⓑ Ⓒ Ⓓ
47. Ⓐ Ⓑ Ⓒ Ⓓ
48. Ⓐ Ⓑ Ⓒ Ⓓ
49. Ⓐ Ⓑ Ⓒ Ⓓ
50. Ⓐ Ⓑ Ⓒ Ⓓ
51. Ⓐ Ⓑ Ⓒ Ⓓ
52. Ⓐ Ⓑ Ⓒ Ⓓ
53. Ⓐ Ⓑ Ⓒ Ⓓ
54. Ⓐ Ⓑ Ⓒ Ⓓ
55. Ⓐ Ⓑ Ⓒ Ⓓ
56. Ⓐ Ⓑ Ⓒ Ⓓ
57. Ⓐ Ⓑ Ⓒ Ⓓ
58. Ⓐ Ⓑ Ⓒ Ⓓ
59. Ⓐ Ⓑ Ⓒ Ⓓ
60. Ⓐ Ⓑ Ⓒ Ⓓ

48. Sperm is formed in the
   (A) Leydig cells
   (B) prostate gland
   (C) seminal vesicles
   (D) seminiferous tubules
   (E) baculum

49. The regulation of the internal environment in animals is referred to as
   (A) equilibrium
   (B) stasis
   (C) homeostasis
   (D) regulation
   (E) feedback

50. Fertilization—the fusion of egg and sperm cell—results in which of the following?
   (A) Embryo
   (B) Zygote
   (C) Gamete
   (D) Ovum
   (E) Follicle

Free-Response Question

1. *Muscle cells are responsible for moving parts of the skeleton by contracting. However, during muscle contraction, none of the muscle cells themselves actually contract.*

   (a) *Describe how a muscle can contract without any of its cells contracting.*

   (b) *Explain the phenomenon of tetanus.*

   (c) *Explain why muscles become "sore" after exercise.*

ANSWERS AND EXPLANATIONS

Multiple-Choice Questions

1. (C) is correct. The only condition listed that is necessary in all organisms that breathe is that moist membranes, also known as respiratory surfaces, are available. The movement of O_2 and CO_2 across the membranes between the environment and the respiratory surface occurs only by diffusion. Respiratory surfaces are generally thin and, since living animal cells must be wet in order to maintain their plasma membranes, these respiratory surfaces must be moist.

2. (C) is correct. The three main stages of development in animals are cleavage, in which a multicellular embryo is created from the zygote through a series of mitotic cell divisions, gastrulation, in which cells migrate and rearrange to form three germ layers, and organogenesis, in which rudimentary organs are formed from the germ layers.

3. (D) is correct. Epinephrine is a hormone that is secreted by the adrenal glands, specifically the adrenal medulla. It functions in raising the blood glucose level, increasing metabolic activities, and constricting blood vessels; all of this prepares the animal for fight or flight response that is elicited in the body during stressful times. The most stressful time listed for most people would be taking an exam while being unprepared.

4. (A) is correct. Erythrocytes are red blood cells, and they transport oxygen around the body. They are the most numerous blood cells, and are small and disc-like in shape. In mammals, erythrocytes have no nuclei. Instead they contain millions of molecules of hemoglobin, which is the iron-containing protein that transports oxygen throughout the body. Hemoglobin can bind four oxygen molecules at once.

5. (D) is correct. The proximal tubule is the site of secretion and reabsorption that substantially changes the content and volume of the filtrate. They secrete hydrogen ions and ammonia to regulate the pH of the filtrate, and they also reabsorb about 90% of the bicarbonate, which is an important buffer.

6. (A) is correct. Salivary amylase is contained in saliva; it is an enzyme that hydrolyzes starch, a glucose polymer found in plants, and glycogen, which is a glucose polymer found in animals. After hydrolysis, smaller polysaccharides and maltose remain.

7. (D) is correct. The adrenal medulla secretes epinephrine and norepinephrine; the regulation of the release of these hormones is controlled by the nervous system, and these hormones act to raise the blood glucose level, increase metabolic activity in the cell, and constrict certain blood vessels—which decreases the amount of blood flow through them.

8. (E) is correct. The pineal gland is small and located near the center of the brain; it secretes the hormone melatonin, which regulates functions related to light and seasons. Most of its functions are related to biological rhythms associated with reproduction, however. It is secreted at night.

9. (C) is correct. The posterior pituitary gland releases two main hormones, oxytocin, which stimulates the contraction of the uterus and mammary gland cells, and antidiuretic hormone (ADH), which promotes the retention of water by the kidney. The actions of the posterior pituitary are regulated by the nervous system, and the water/salt balance in the body.

10. (A) is correct. The ovaries secrete hormones called estrogens, which stimulate the growth of the uterine lining and promote the development of secondary sex characteristics in females. They are regulated by two other hormones, FSH and LH.

11. (B) is correct. The thyroid gland releases the hormones triiodothyronine and thyroxine, which stimulate and maintain metabolic processes and lower the blood calcium level, respectively. The thyroid gland secretions are regulated by TSH and by the level of calcium in the blood.

12. (B) is correct. Blood is a connective tissue. It functions very differently from the other connective tissues, but it has an extensive extracellular matrix that is the criteria for being connective tissue. The matrix is plasma, which consists of water, salts, and dissolved proteins.

13. (A) is correct. The three types of muscle in the body are skeletal muscle (responsible for voluntary movements), cardiac muscle (which forms the contractile wall of the heart), and smooth muscle (found in the walls of the digestive tract, bladder, arteries, and other internal organs).

14. (D) is correct. The traditional example of a negative feedback system is the thermostat example. In the body, one very prominent example of negative feedback is the regulation of our body temperature at about 37° Celsius. A section of the brain is responsible for keeping track of the temperature of the blood, and if the blood is too warm, for example, it tells the sweat glands to increase production.

■ **15. (B) is correct.** The fat soluble vitamins are vitamins A, D, E, and K. Vitamin A is used in pigments in the eye; vitamin D helps in calcium absorption and bone formation; vitamin E has an unknown function; and vitamin K is required for blood clotting.

■ **16. (C) is correct.** Herbivores are animals that eat only autotrophs, and autotrophs are plants. Some examples of herbivores are gorillas and cows. Carnivores eat other animals, and omnivores eat animals as well as plant or algae.

■ **17. (C) is correct.** The four stages of food processing are ingestion (the act of eating), digestion (the process by which food is broken down into small particles the body can absorb), absorption (the uptake of nutrients by the body), and elimination (the release of undigested material).

■ **18. (E) is correct.** Hydra are simple animals that contain a gastrovascular cavity, which is a pouch that functions in both digestion and the distribution of nutrients throughout the body. The gastrovascular cavity's single opening acts as both mouth and anus.

■ **19. (E) is correct.** The liver is responsible for the production of bile, which contains no digestive enzymes but does contain bile salts (which act as detergents or emulsifying agents that help digest fats). Bile also contains pigments that are the byproducts of red blood cells destroyed in the liver. These are eliminated from the body along with the feces.

■ **20. (D) is correct.** The digestion of carbohydrates, like starch and glycogen, begins in the mouth through the action of salivary amylase. In the small intestine, pancreatic amylases hydrolyze starch, glycogen, and smaller polysaccharides into disaccharides. There maltase completes the digestion of maltose by splitting it into glucose.

■ **21. (B) is correct.** Pepsin is an enzyme that is present in the gastric juice of the stomach. It begins the hydrolysis of proteins by breaking peptide bonds between adjacent amino acids and by cleaving proteins into smaller polypeptides. The digestion of proteins continues in the small intestine by the enzymes trypsin and chymotrypsin.

■ **22. (D) is correct.** The large intestine, also known as the colon, is responsible for recovering water that was in the alimentary canal to act as a solvent for the material being digested. It is also responsible for compacting the wastes into feces, which are stored in the rectum and then excreted.

■ **23. (D) is correct.** Insects, other arthropods, and many mollusks have hemolymph which circulates through their bodies. In these animals, blood and interstitial fluid are mixed together, and one or more hearts pump this fluid throughout an interconnected network of sinuses (spaces that surround the organs).

■ **24. (A) is correct.** In an open circulatory system, which exists in insects and other arthropods, the blood bathes the organs directly. In closed circulatory systems, blood is contained in vessels and is separate from the interstitial fluid.

In closed circulatory systems, one or more hearts pump blood into vessels that branch and feed blood through the vessels.

25. **(C) is correct.** An artery is a kind of blood vessel that carries blood away from the heart, branching into arterioles and eventually into capillary beds. The capillary beds then converge into venules, which converge further into veins, which return blood to the heart.

26. **(D) is correct.** The role of the sinoatrial node (SA node) or pacemaker is to control the rate and timing of the contraction of heart muscles. It generates nerve impulses just like the ones that occur in nerve cells, and the impulses spread rapidly through the walls of the atria, making them contract in unison.

27. **(C) is correct.** The lymphatic system collects fluid and proteins lost during regular circulation and returns them to the blood. This system is composed of a network of lymph vessels throughout the body, with periodic lymph nodes, which are the sites at which lymph is filtered and viruses and bacteria are attacked.

28. **(E) is correct.** All of the answers listed except free amino acids are constituents of blood. White blood cells and leukocytes are the same thing, and red blood cells are also called erythrocytes. Amino acids would be found free in the digestive tract, but not in the blood vessels.

29. **(C) is correct.** Erythrocytes, leukocytes, and platelets all develop from stem cells in the red marrow of bones—primarily in the ribs, vertebrae, breastbone, and pelvis. The cells that develop into blood cells have the potential to develop into any type of blood cell; they are called pluripotent cells.

30. **(C) is correct.** Insects have a tracheal system, which is made up of air tubes that branch throughout the body. The large tubes are called trachea, and they open to the outside, while the smallest branches reach the surface of every cell, where gas exchange takes place.

31. **(D) is correct.** Carbon dioxide is most commonly transported in the blood in the form of bicarbonate—it reacts with water to form carbonic acid, and then a hydrogen dissociates from carbonic acid to produce bicarbonate. Less commonly, carbon dioxide is transported by hemoglobin, or transported in solution in the blood.

32. **(E) is correct.** All of the answers listed—except phagocytes—are examples of first-line barriers to infection by infecting agents that might attack the body. Phagocytosis constitutes the body's nonspecific internal mechanism for defending itself against infectious agents; it is the process by which invading organisms are ingested and destroyed by white blood cells.

33. **(B) is correct.** When a lymphocyte is activated by an antigen, it is stimulated to divide and differentiate, and it forms two clones. One clone is of effector cells that combat the antigen. One clone is of memory cells that stay in the circulation, recognize the antigen if it infects the body in the future, and launch an attack against it.

34. **(A) is correct.** All of the answers except *A* constitute methods animals have for exchanging heat with the environment. Conduction is the transfer of heat between objects that are in direct contact. Convection is the transfer of heat by

the movement of air past a surface. Radiation is the emission of electromagnetic waves by warm objects. Evaporation is heat loss through the loss of molecules as gas.

35. (B) is correct. Ectotherms are animals that have such low metabolic rates that the amount of heat they generate will not influence their body temperature. Their internal temperature is therefore determined by their environment. Endotherms have high metabolic rates, and this makes their bodies quite a bit warmer than the external environment.

36. (D) is correct. Malpighian tubules are organs that remove the nitrogenous wastes of insects and other arthropods. They open into the digestive tract and dead-end at tips that are submerged in hemolymph. The tubules have an epithelial lining that secretes solutes into the lumen of the tubule, and water follows the solutes into the tubule by osmosis.

37. (D) is correct. The nephron, which is the functional unit of the kidney, is composed of a long tubule and the glomerulus, which is a ball of capillaries. The Bowman's capsule surrounds the glomerulus. The blood in the glomerulus is forced into the Bowman's capsule by blood pressure, and this process acts to filter the blood.

38. (B) is correct. Rod cells and cone cells are the two types of photoreceptors in the eye. They are contained in the retina and account for 70% of all the sensory receptors in the body. Cone cells can distinguish colors in daylight, whereas rods are sensitive to light but cannot distinguish colors.

39. (D) is correct. The cochlea is part of the inner ear that is involved in hearing. It is a coiled organ that has two large chambers—a vestibular canal and a lower tympanic canal, which are separated by a cochlear duct. The floor of the cochlear duct is home to the organ of Corti, which contains the receptors of the ear, hair cells.

40. (A) is correct. The vitreous humor is jellylike and fills the posterior cavity of the eye, constituting most of the eye's volume. The aqueous humor fills the anterior cavity of the eye and is clear and watery.

41. (C) is correct. On one end, the Eustachian tube connects to the middle ear, and on the other, the Eustachian tube connects with the pharynx. This enables it to equalize the pressure between the middle ear and the atmosphere.

42. (E) is correct. Taste buds are modified epithelial cells that act as receptors for taste. Most taste buds are on the surface of the tongue and mouth. Sweet, sour, salty, and bitter are taste perceptions detected by taste buds.

43. (D) is correct. The sliding-filament model of muscle contraction states that the thin and thick filaments do not shrink during muscle contraction. Instead the filaments slide past each other so that the degree of their overlap increases; this sliding is based on the interactions of actin and myosin molecules that make up the filaments.

44. (B) is correct. There are three successive stages of development that follow fertilization. The first is cleavage, which is rapid cell division that produces a mass of new cells that share the cytoplasm of the original cell. The new cells all have their own nuclei and are called blastomeres. The second stage is gastrulation, and the third is organogenesis.

45. **(B) is correct.** The reflex arc is the simplest type of nerve circuit (automatic response), and it requires just two types of cells. A sensory neuron receives information from a receptor and passes it to the spinal cord and then to a motor neuron, which signals an effector cell to respond to the stimulus.

46. **(C) is correct.** Neurotransmitters are excreted by the synaptic vesicles and act as intercellular messengers, transmitting the nerve impulse from one neuron to the next. A single postsynaptic neuron can receive signals from many neurons that secrete different neurotransmitters.

47. **(A) is correct.** Each of the two ovaries in the female body contains many follicles. Follicles are composed of an egg cell surrounded by one or two layers of follicle cells; these serve to nourish and protect the cell. All of a woman's follicles are formed before her birth.

48. **(D) is correct.** Sperm is produced in the seminiferous tubules, which are coiled tightly in the testes and surrounded by connective tissue. Production of sperm cannot take place at the high temperature of the body, so the testes are held in the scrotum of the male, outside the abdominal pelvic cavity, where it is about 2 degrees cooler.

49. **(C) is correct.** Homeostasis is the ability of many animals to regulate their internal environment. They do this through thermoregulation, which is the maintenance of internal temperature in a certain range, and osmoregulation, which is the regulation of solute balance within certain parameters.

50. **(B) is correct.** Fertilization is the fusion of egg cell (ovum) and the sperm cell, and it results in the formation of a zygote. The zygote is diploid, whereas the egg cell and the sperm cell are both the products of meiosis, and so are haploid.

Free-Response Question

1. (a) Skeletal muscle is fibrous, and each fiber is a single long cell. The fibers are composed of myofibrils, which are in turn composed of two kinds of myofilaments. These are thin filaments—which are made up of two actin strands and one regulatory protein strand, coiled—and thick filaments made up of myosin molecules. The sarcomere is the basic contracting unit of the muscle; during muscle contraction, the length of the sarcomere decreases. The sliding-filament model of muscle contraction states that the thick and thin filaments slide past each other horizontally, due to the interactions of actin and myosin.

   The myosin molecules look like golf clubs arranged horizontally in a group, with the head of the golf club pointing up. This head region is the center of the reactions that take place during muscle contraction. Myosin binds ATP and hydrolyzes it into ADP, and its structure is changed in the process, which causes it to bind to a specific site on actin and form a cross-bridge. Myosin then releases the stored energy and relaxes to its normal conformation. This changes the angle of attachment of the myosin head relative to its tail. As myosin bends inward upon itself, tension increases on the actin filament, and the filament is pulled toward the middle of the sarcomere.

(b) In the transmission of action potentials through muscle cells, in response to a nerve impulse, a single action potential will cause an increase in tension in the muscle cell, and if a second action potential arrives within a certain short period of time, the response will be greater; the two responses are summed. If a muscle cell receives action potentials from many nerve cells surrounding it, these too will be summed, and the level of tension will depend on how quickly the action potentials follow one another. If the rate of stimulation is sufficiently high, the muscle twitches will blur, and tetanus will result.

(c) When oxygen is scarce, as in situations where a person is taking part in strenuous exercise, human muscle cells switch to lactic acid fermentation (which normally undergoes regular aerobic cell respiration) to produce ATP. In lactic acid fermentation, pyruvate is reduced by NADH to form lactate with no release of CO_2. The lactate that accumulates as a result of this reaction can cause muscle fatigue and pain.

This response uses the following key terms in context, showing the writer's knowledge of their meanings and relatedness:

skeletal muscle	*cross-bridge*
myofibrils	*action potential*
myofilaments	*summation*
thin filaments	*tetanus*
actin strands	*aerobic cell respiration*
thick filaments	*lactic acid fermentation*
myosin	*pyruvate*
sarcomere	*lactate*
sliding-filament model	

It also shows knowledge of how these important biological processes take place: how muscle cells contract and ultimately cause bones to move; how tetanus is reached; and how and why strenuous exercise produces muscle pain.

CHAPTER 8

Ecology

An Introduction to Ecology and the Biosphere: The Scope of Ecology

> The **abiotic components** of an environment are the nonliving, chemical, and physical components; the **biotic components** are the living components of an environment.

Factors Affecting the Distribution of Organisms

> ▎ **Biogeography** is the study of the geographic distribution of species in the past and today.
>
> ▎ The **dispersal** of organisms refers to their global geographic distribution. Some **abiotic factors** that affect the distribution of organisms are temperature, water, sunlight, wind, and the composition of the rocks and soil.
>
> ▎ The components that make up the **climate** in a certain location are temperature, water, light, and wind.
>
> ▎ **Biomes** are the major types of ecosystems that occupy very broad geographic regions. Examples include coniferous forests, deserts, and grasslands.
>
> ▎ The **tropics** receive the greatest amount of sunlight annually and the least amount of solar variation through the course of the year.
>
> ▎ **Microclimates** are small-scale environmental variations , like under a log.

Aquatic and Terrestrial Biomes

> ▎ **Aquatic biomes** take up the largest part of the biosphere, because water covers roughly 75% of Earth's surface. These biomes are classified into **freshwater biomes** and **marine biomes**.
>
> ▎ Aquatic biomes display vertical stratification; there is a **photic zone** (in which there is enough light for photosynthesis to occur) and an **aphotic zone** (where very little light penetrates).
>
> ▎ **Thermoclines** are narrow layers of fast temperature change that separate a warm upper layer of water and cold deeper waters.
>
> ▎ The **benthic zone** is located at the bottom layer of all aquatic biomes, and it is made up of sand and organic and inorganic sediments (including **detritus**, dead organic matter). Organisms that live in the benthic zone are called **benthos**.
>
> ▎ The **two types of freshwater biomes** are standing bodies of water, such as lakes, and moving bodies of water, such as streams.
>
> ▎ In lakes, organism communities are distributed according to the water's depth. The **littoral zone** (well-lit shallow waters near the shore) contains rooted and

floating aquatic plants, whereas the **limnetic zone** (well-lit open surface waters farther from shore) are occupied by phytoplankton.

▮ The **profundal zone** is the deep aphotic region in a lake. This is where the remains of small organisms that live in the limnetic zone sink on their way to the benthic zone.

▮ **Oligotrophic lakes** are deep lakes that lack nutrients and contain sparse phytoplankton; **eutrophic lakes** are shallower, and they have higher nutrient content with a high concentration of phytoplankton.

▮ **Wetlands** are areas covered with water deep enough to support aquatic plants.

▮ **Estuaries** are areas where freshwater streams or rivers merge with the ocean.

▮ In marine (saltwater) communities, the zone where land meets water is called the **intertidal zone**, and beyond the intertidal zone is the **neritic zone**—the shallow water over the continental shelves. Past the continental shelves is the **oceanic zone**. Any open water is called the **pelagic zone**, and—as in freshwater systems—the **benthic zone** lies at the bottom of the water at the ocean's floor. The **abyssal zone** refers to very deep benthic communities.

▮ **Coral reefs** inhabit the neritic zone. A coral reef is a biome created by a group of cdniarians that secrete hard calcium carbonate shells that vary in shape and support the growth of other corals, sponges, and algae.

▮ **Canopy** refers to the upper layers of trees in a forest.

▮ **Permafrost** is the permanently frozen stratum that exists in some biomes.

▮ Some major terrestrial biomes are listed below.

1. **Savannas**—Here the dominant herbivores are insects, such as ants and termites, and the dominant plants are grasses and some trees. Fire is a dominant abiotic factor, and many plants are adapted for fire. Plant growth is quite substantial during the rainy season, but large grazing mammals must migrate during regular seasons of drought.

2. **Desert** is marked by sparse rainfall, and desert plants and animals are adapted to conserve and store water. Deserts contain many CAM plants and plants with adaptations that prevent animals from consuming them, like cacti. Temperature (either hot or cold) is usually extreme.

3. **Chaparral** is dominated by dense, spiny, evergreen shrubs. These are coastal areas with mild rainy winters and long, hot, dry summers. Plants are adapted to fires.

4. **Temperate grassland** is marked by seasonal drought with occasional fires, and by large grazing mammals. All of these factors prevent the significant growth of trees. Grassland soil is rich in nutrients, and these areas are good for agriculture.

5. **Temperate deciduous forest** is marked by dense strands of deciduous trees that require sufficient moisture. These forests are more open than (and not as tall as) rain forests. They are stratified—the top layer contains one or two strata of trees; beneath that are shrubs; and under that is an herbaceous stratum. These trees drop their leaves in fall, and many enter hibernation. Many birds migrate to warmer climates.

6. **Coniferous forest** is dominated by cone-bearing trees like pine, spruce, and fir. The conical shape of conifers prevents much snowfall from accumulating on—and breaking—these trees' branches.

7. **Tundra** possesses permafrost, very cold temperatures, and high winds, and tundra supports no trees or tall plants. Tundra receives very little rain and accounts for 20% of the earth's terrestrial surface.

8. **Tropical rainforest** has pronounced vertical stratification. Trees in a canopy are at the top, and they are dense enough that little light breaks through. These forests are marked by epiphytes, which are plants that grow on other plants instead of the soil. Rainfall is varied.

Behavioral Biology: Introduction to Behavior and Behavioral Ecology

▐ **Behavior** is what an animal does and how it does it. Behavior is a result of genetic and environmental factors. **Innate behavior** is that behavior which is developmentally fixed. A **fixed action pattern** is a sequence of behaviors that is largely unchangeable and usually carried to completion once it is initiated. Fixed action patterns are triggered by sign stimuli.

▐ **Behavioral ecology** is a scientific field of study that sees behavior as an evolutionary adaptation to ecological conditions.

▐ **Ethnology** is the study of animal behavior.

Learning

▐ **Learning** is defined as the modification of behavior that results from specific experiences.

▐ **Habituation** is a simple type of learning that involves a loss of responsiveness to stimuli that convey little or no information.

▐ **Imprinting** is learning that is limited to a certain time period in an organism's life, and it is generally irreversible.

▐ **Associative learning** is the ability of some animals to learn to associate one stimulus with another. **Classical conditioning** involves learning to associate a certain stimulus with a reward or a punishment. **Operant conditioning** occurs as an animal learns to associate one of its behaviors with a reward or punishment.

Animal Cognition

▐ **Cognition** is the ability of an animal's nervous system to perceive, store, process, and use information from sensory receptors.

▐ A **kinesis** is a simple change in activity in respond to a stimulus, whereas a taxis is an automatic movement toward or away a stimulus.

▐ A **cognitive map** is an internal representation of spatial relationships among objects in an animal's surroundings.

Social Behavior and Sociobiology

▐ **Social behavior** is any kind of interaction between two animals, and **sociobiology** applies evolutionary theory to the study of social behavior.

▐ **Agonistic behavior** refers to a contest that involves threatening and submissive behavior over a resource. Usually agonistic behavior involves **ritual**, or the use of symbolic activity.

▐ A **dominance hierarchy** is a "pecking order" in a group.

- **Territory** refers to an area that an individual defends, usually excluding other members of its species.
- **Courtship** is a set of behavior patterns that lead to copulation. **Parental investment** refers to the time and resources an individual spends to produce and nurture offspring.
- A **signal** is a behavior that causes a change in the behavior of another individual. **Pheromones** are chemical signals that are emitted by animals.
- **Altruism** refers to events in which animals behave in ways that reduce their individual fitness and that increase the fitness of the recipient of the behavior. **Kin selection** is natural selection that favors altruistic behavior by enhancing the success of relatives.

Population Ecology: Characteristics of Populations

A **population** is a group of individuals of a single species that occupies a certain geographic area.

Population Growth and Population-Limiting Factors

- **Exponential population growth** refers to population growth that exists under ideal conditions.
- The **carrying capacity** of a population is defined as the maximum population size that a certain environment can support at a particular time with no degradation of the habitat.
- When a death rate rises as population density rises, the death rate is said to be **density dependent**. When a death rate does not change with increase in population density, it is said to be **density independent**.
- Some populations have regular **cycles** of boom and bust.

Community Ecology: What Is a Community?

A **community** is an assemblage of populations in an area or habitat.

Interspecific Interactions and Community Structure

- **Interspecific competitions** for resources occur when resources are in short supply.
- The **competitive exclusion principle** states that when two species are vying for a resource, eventually the one with the slight reproductive advantage will eliminate the other.
- An organism's ecological **niche** is the sum total of biotic and abiotic resources in its environment.
- **Predation** is an interaction between two species in which one species (the **predator**) eats the other species (the **prey**). One defense that animals have against predators is **cryptic coloration**, in which the animal is camouflaged by its coloring. Another defense is **aposematic coloration**, in which a poisonous animal is brightly colored as a warning to other animals.
- **Batesian mimicry** refers to a situation in which a non-poisonous animal has evolved to mimic the coloration of a poisonous animal. In **Müllerian**

mimicry, two poisonous species evolve to resemble each other, ostensibly so that predators will learn to avoid them more quickly.

❚ The **trophic structure** of a community refers to the feeding relationships among the organisms. **Trophic levels** are the links in the trophic structure of a community.

❚ The transfer of food energy from plants through herbivores through carnivores through decomposers is referred to as a **food chain**. **Food webs** are food chains hooked together.

❚ **Dominant species** in a community have the highest **biomass** (the sum weight of all the members of a population).

❚ **Keystone species** are the most abundant species in a community.

Disturbance and Community Structure

❚ **Ecological succession** refers to transitions in species composition in a certain area over ecological time.

❚ In **primary succession**, plants and animals invade a region that was lifeless.

❚ **Secondary succession** occurs when an existing community has been cleared by a disturbance that leaves the soil intact.

Biogeographic Factors Affecting the Biodiversity of Communities

The **biodiversity** of a community (its species diversity) is determined by its size and geographic location.

Ecosystems: The Ecosystem Approach to Ecology

❚ **Primary producers** in an ecosystem are the **autotrophs**, and they support all the others in the ecosystem.

❚ Organisms that are in trophic levels above primary producers are **heterotrophs**.

❚ Herbivores eat primary producers and are called **primary consumers**.

❚ Carnivores that eat herbivores are called **secondary consumers**, and carnivores that eat other carnivores are called **tertiary consumers**.

❚ **Detritivores**, or **decomposers**, get their energy from detritus, which is nonliving organic material such as the remains of dead organisms, feces, dead leaves, and wood.

The Cycling of Chemical Elements in Ecosystems

❚ **Biogeochemical cycles** are nutrient cycles that contain both biotic and abiotic components.

❚ Most of the earth's nitrogen is in the form of N_2, which is unusable by plants.

❚ **Nitrogen fixation** occurs when organisms convert N_2 to usable nitrogenous compounds.

❚ **Nitrification** is the process by which ammonium (NH_4^+) is oxidized to nitrite and then nitrate (NO_3^-).

❚ **Denitrification** is the process by which some bacteria can get the oxygen they need for metabolism from nitrate rather than from O_2 under certain conditions.

❚ The decomposition of nitrogen back to ammonium is called **ammonification**.

Human Impact on Ecosystems and the Biosphere

▌ **Acid precipitation** is defined as rain, snow, or fog with a pH less than 5.6.

▌ In **biological magnification**, toxins become more concentrated in successive trophic levels of a food web.

▌ The **greenhouse effect** refers to the rise in atmospheric CO_2 concentration that causes solar radiation to be re-reflected back to the earth.

▌ Because of the burning of fossil fuels, CO_2 levels have been steadily increasing. One effect of this increase is that **earth is being warmed significantly**.

▌ The **ozone layer** reduces the amount of penetration of UV radiation from the sun through the atmosphere. Chlorine-containing compounds used by humans are eroding the ozone layer, and this could have disastrous effects in the future.

For Additional Review:

Consider the possible effects of global warming. What would happen if the earth's temperature rose by ten degrees? How would this affect the biotic and abiotic components of the earth?

Multiple-Choice Questions

1. All of the following statements about the earth's ozone layer are false EXCEPT
   (A) it is composed of O_2
   (B) it amplifies the amount of ultraviolet radiation that reaches the earth
   (C) it is thinning as a result of widespread use of certain chlorine-containing compounds
   (D) it is thickening as a result of widespread use of certain chlorine-containing compounds
   (E) it allows green light in but screens out red light

2. Which of the following is the major primary producer in a grassland ecosystem?
   (A) Lion
   (B) Gazelle
   (C) Grass
   (D) Snake
   (E) Diatom

3. The carrying capacity of a population is defined as
   (A) the amount of time the parents in the population spend rearing and nurturing their offspring

   (B) the maximum population size that a certain environment can support at a particular time
   (C) the amount of vegetation that a certain geographic area can support
   (D) the number of different types of species a biome can support
   (E) the number of different genes a population can carry at a particular time

4. Which of the following is the term used to describe major types of ecosystems that occupy broad geographic regions?
   (A) Biome
   (B) Community
   (C) Chaparral
   (D) Trophic level
   (E) Biosphere

5. A lake that is full of nutrients and that supports a vast array of algae is said to be
   (A) oligotrophic
   (B) profundal
   (C) littoral
   (D) eutrophic
   (E) limnetic

6. Which of the following best describes an estuary?
   (A) An area that is periodically flooded so its soil is consistently damp
   (B) An area where a river changes course after being diverted from its original course by an obstacle
   (C) The area where a freshwater river merges with the ocean
   (D) The area where a mass of cold water and a mass of warm water meet in the pelagic zone
   (E) An outshoot of land that extends into the ocean

7. Which of the following is the term that refers to the layer of inorganic and organic nutrients that layers the ocean floor?
   (A) Littoral zone
   (B) Limnetic zone
   (C) Profundal zone
   (D) Benthic zone
   (E) Photic zone

Directions: The group of questions below consists of five lettered headings followed by a list of numbered phrases or sentences. For each numbered phrase or sentence select the one heading that is most closely related to it and fill in the corresponding oval on the answer sheet. Each heading may be used once, more than once, or not at all.

Questions 8–12
   (A) Temperate grassland
   (B) Tropical rainforest
   (C) Temperate deciduous forest
   (D) Tundra
   (E) Desert

8. Characterized by permafrost and few large plants

9. Characterized by epiphytes and significant canopy

10. Characterized by trees that lose their leaves in the fall and home to migrating birds

11. Characterized by frequent fires and nutrient-rich soil

12. Characterized by sparse rainfall and fluctuating temperatures

13. Fixed action patterns are instigated by which of the following?
   (A) Mating behavior
   (B) Ritual behavior
   (C) Innate stimulus
   (D) Sign stimulus
   (E) Action potential

14. One morning, a woman who usually feeds her two cats in the morning passes by the food bowl without putting food in it. The cats usually run over to the bowl as she approaches it, but after four mornings of her passing the bowl without putting food in it, the cats no longer run over to the bowl. This is an example of
   (A) maturation
   (B) imprinting
   (C) habituation
   (D) foraging
   (E) sensitivity

15. Pavlov's dogs learned to salivate when they heard the ring of a particular bell; this is an example of
    (A) classical conditioning
    (B) operant conditioning
    (C) sensitivity
    (D) imprinting
    (E) maturation

16. The phenomenon in which young ducks follow their mother in a line is a result of which of the following?
    (A) Habituation
    (B) Imprinting
    (C) Maturation
    (D) Foraging
    (E) Conditioning

17. Altruism exists in populations because
    (A) species within a community grow to love each other
    (B) it can result in the passing on of the altruistic member's genes
    (C) it can result in the overall success of the community
    (D) it can result in a bond between the altruistic member and the recipient who receives altruism, and the recipient might later reciprocate the altruism
    (E) it can result in the maximizing of the altruistic member's genetic representation in a population, if the altruistic member's behavior is directed toward a close relative

18. A bacterial colony that exists in an environment displaying ideal conditions will undergo
    (A) logistic growth
    (B) explosive growth
    (C) hyperactive growth
    (D) exponential growth
    (E) unbounded growth

19. A species' specific use of the biotic and abiotic factors in an environment is collectively called the species'
    (A) habitat
    (B) trophic level
    (C) niche
    (D) placement
    (E) partitioning

20. In which type of camouflaging does a non-toxic animal mimic the appearance of a toxic animal?
    (A) Müllerian mimicry
    (B) Cryptic coloration
    (C) Aposematic coloration
    (D) Batesian mimicry
    (E) Parasitoidism

21. The dominant species in a community is the one that
    (A) has the greatest number of individuals
    (B) is at the top of the food chain
    (C) has the largest biomass
    (D) eats all other members of the community
    (E) bears the most offspring in each mating

22. Several food chains are hooked together in a
    (A) trophic structure
    (B) food web
    (C) energy transfer web
    (D) community
    (E) population

23. A fire cleared a large area of forest in Yellowstone National Park in the 1980s. When the first plants pioneered this area, this was an example of
    (A) primary succession
    (B) secondary succession
    (C) pioneering
    (D) a keystone species
    (E) the top-down model

24. In the nitrogen cycle, the process by which organic nitrogen is decomposed back to ammonium is known as
(A) ammonification
(B) denitrificaton
(C) nitrogen-fixation
(D) nitrogen cycling
(E) nitrogenation

25. The process in which CO_2 in the atmosphere intercepts and absorbs reflected infrared radiation and re-reflects it back to the earth is known as
(A) global warming
(B) atmospheric insulation
(C) stratospheric insulation
(D) biological magnification
(E) the greenhouse effect

Free-Response Question

1. All of the organisms in a community are interrelated by the abiotic and biotic resources they use in the course of their lives.

▌ Describe the relationships that exist among a hawk, a mouse, a plant, and the earth in a particular ecosystem.

▌ As unlikely as it may seem, biotic components of an environment do influence the abiotic components of an environment. Give two examples of this influence.

ANSWERS AND EXPLANATIONS

Multiple-Choice Questions

▌ **1. (C) is correct.** The ozone layer is located in the stratosphere and surrounds the earth. It is composed of O_3, and it absorbs UV radiation, preventing it from reaching the organisms in the biosphere. The ozone layer has been thinning since about 1975. The destruction of the ozone layer has been attributed to the widespread use of chloroflourocarbons.

▌ **2. (C) is correct.** In a grassland, grass constitutes the primary producer. The primary producer is the one that traps the energy of sunlight and turns it into chemical energy. Primary consumers consume primary producers; secondary consumers eat herbivores; and tertiary consumers eat carnivores.

▌ **3. (B) is correct.** The carrying capacity of a population is defined as the maximum population size a particular environment can support at a particular time with no degradation of the habitat. It is fixed at certain times, but it varies over the course of time with the amount of resources that exist in an environment.

▌ **4. (A) is correct.** Biomes are major types of ecosystems that occupy broad geographic regions. Some examples of biomes are coniferous forests, deserts, grasslands, and tropical forests.

▌ **5. (D) is correct.** Lakes are classified according to how much organic matter they produce. Oligotrophic lakes are deep and generally poor in nutrients, and therefore they have relatively little phytoplankton, whereas eutrophic lakes are usually shallower and have greater nutrient content, which allows the growth of more phytoplankton.

6. (C) is correct. An estuary is an area where a running freshwater source, such as a stream or river, meets the ocean. Often estuaries are bordered by large areas of coastal wetlands, and salinity varies with location within them, as well as with the rise and fall of the ocean tides. Estuaries are one of the most biologically productive biomes, and they are also home to many of the fish and other animals that humans consume.

7. (D) is correct. The benthic zone refers to the lowest of all the biotic zones of any aquatic biome. This bottommost layer is composed of sand and inorganic and organic sediments, and it is home to communities of organisms called benthos. These organisms derive nutrients from dead organic matter, such as dead fish and other ocean life, that falls down from the zones above.

8. (D) is correct. Tundra is characterized by having permafrost (which is a permanently frozen topsoil), very cold temperatures, and high winds. These factors of the climate are what prevent tall plants from growing in the tundra. Tundra generally does not receive much rainfall throughout the year, and what rain does fall cannot soak into the earth because of the permafrost.

9. (B) is correct. Tropical rainforests generally have thick canopies that prevent much sun from filtering through. This means that in breaks in the canopy, other plants grow quickly to compete for sunlight. Tropical rainforests are home to epiphytes, and rainfall is frequent.

10. (C) is correct. Temperate deciduous forests are characterized by dense populations of deciduous trees, which drop their leaves in the fall when the weather turns cold. They do this because they won't be able to get enough water to support their leaves, when the groundwater is frozen. These forests are also home to birds that migrate to warmer climates when the temperature drops.

11. (A) is correct. Temperate grasslands are characterized by having thick grass, seasonal drought, occasional fires, and large grazing animals. Their soil is generally rich with nutrients, so they are good for agriculture. Most of the temperate grassland in the United States is used today for agriculture.

12. (E) is correct. Deserts experience very little rainfall, so they are home to many animals that have adaptations for storing and saving water. Deserts are marked by drastic temperature fluctuation; they can be very hot in the day but freezing at night. Many desert plants rely on CAM photosynthesis.

13. (D) is correct. Fixed action patterns are a sequence of behavioral acts that are virtually unchangeable and usually carried through to completion once they are initiated. Sign stimuli trigger fixed action patterns, and often sign stimuli are things like a feature of another animal, some aspect of its appearance, or some other event.

14. (C) is correct. Habituation is one type of learning. Learning is defined as the ability of an animal to modify its behavior as a result of specific experiences, and habituation is a very simple form of learning, in which there is a loss of responsiveness to stimuli that convey very little or no information.

15. (A) is correct. Classical conditioning is a form of associative learning (the ability of animals to learn to associate one stimulus with another). It specifically refers to an animal's ability to associate an arbitrary stimulus with a reward or a punishment.

16. (B) is correct. Imprinting is a form of learning that occurs once in a life, during a sensitive period. Imprinting is generally irreversible, and the sensitive period is a limited phase in the animal's development when the learning of particular behavior can take place.

17. (E) is correct. Altruism occurs in populations because if parents sacrifice their own well-being for that of their offspring, this increases their fitness by better insuring that the genes that they passed on will make it to the next generation. Likewise, helping other close relatives increases the chances that they will survive to pass on genes that are shared between them and the altruistic member.

18. (D) is correct. A bacterial colony growing where it has limitless nutrients, and other ideal conditions, will experience what is called exponential growth. In exponential growth, all members are free to reproduce at their physiological capacity. Populations do not stop growing until they reach their carrying capacity.

19. (C) is correct. A species' ecological niche is defined as the sum of its use of the abiotic and biotic factors in an environment. For instance, a particular bird's niche refers to the food it consumes, what type of trees it builds its nest in, the time of day it is active, and what climate it lives in.

20. (D) is correct. In Batesian mimicry, a harmless or palatable animal can evolve to have the same markings and/or colorings as a harmful or unpalatable animal, and in this way can escape predation.

21. (C) is correct. The dominant species in a community has the highest biomass, or sum weight of all of the members of a population. The dominant species are also hypothesized to be the most competitive in exploiting the resources in an ecosystem.

22. (B) is correct. The trophic structure of a community describes all of the feeding relationships that exist among organisms in the community. It can be divided into specific food chains, which are composed of primary produces, primary consumers, secondary consumers, and tertiary consumers. Food webs are made up of trophic levels and can be combined into more complex food webs.

23. (B) is correct. Secondary succession refers to a situation in which a community has been cleared by a disturbance of some kind, but the soil is left intact. The area will begin to return to its original state through the process of plants invading the area and recolonizing.

24. (A) is correct. In the nitrogen cycle, bacteria oxidize ammonium to nitrite. The bacteria then release the nitrate to be used by plants and converted to organic forms, like amino acids in proteins. Later, the decomposition of organic nitrogen back to ammonium, which is carried out by decomposers, is referred to as ammonification.

25. (E) is correct. The greenhouse effect is the process by which carbon dioxide and water vapors in the atmosphere intercept reflected infrared radiation from the sun and re-reflect it to the earth. Global warming is the process by which the amount of carbon dioxide in the atmosphere is increasing because of humans' combustion of fossil fuels.

1. The hawk, mouse, and plant in this particular ecosystem are related by the passage of energy through them. Together they comprise a food chain—the mouse is a primary consumer, and he consumes the plant, which is a primary producer (the plant is an autotroph—capable of trapping the energy of the sun and converting it into chemical energy in the form of carbohydrates). The hawk then is a predator of the mouse—and a secondary consumer. Secondary consumers eat herbivores. All of these animals are dependent on the soil in which the plant has grown, because the soil is capable of providing the plant with nutrients that it could not get on its own. The plant needs a variety of organic elements to produce carbohydrate, but it also needs mineral elements, and nitrogen is a very important one (it needs nitrogen to make proteins and nucleic acids). Plants are incapable of using nitrogen in its usual form when found in soil (ammonium), and so plants must rely on nitrogen-fixing bacteria to convert it to a useable form, nitrate.

   One example in which the biotic factors of the biosphere are impacting the abiotic factors is seen in the case of global warming. We rely on the greenhouse effect (in which atmospheric carbon dioxide acts as an insulator of sorts, trapping infrared radiation from the sun and re-reflecting it back to us) to help maintain the hospitable temperature of the earth. Yet, due to the use of fossil fuels—beginning during the Industrial Revolution—the concentration of carbon dioxide in the atmosphere has increased significantly, and this has led to an increase in global temperatures.

   Another way in which humans (a biotic factor of the biosphere) are impacting abiotic processes is in the ozone layer. We are protected from ultraviolet radiation from the sun by a protective layer of ozone which surrounds the earth. However, the ozone layer is being degraded by humans' use of chloroflourocarbons, which are chemicals used in refrigeration and other industrial processes.

   This response uses the following key terms in context, showing the writer's knowledge of their meanings and relatedness:

ecosystem	*predator*
primary producer	*nitrogen fixation*
primary consumer	*herbivore*
secondary consumer	*global warming*
food chain	*ozone layer*
autotroph	*chloroflourocarbons*

   This response also shows knowledge of the following important biological processes—food chains and the interaction of organisms in a community, global warming, and the depletion of the ozone layer.

Part III

Sample Tests with Answers and Explanations

On the following pages are two sample examinations that approximate the actual AP Biology Examination in format, types of questions, and content. Set aside three hours to take each test. To best prepare yourself for actual AP exam conditions, use only the allowed amounts of time for Section I and Section II.

Practice Test 1

Biology
Section 1

Time—1 hour and 30 minutes

Directions: Each of the questions or incomplete statements below is followed by five suggested answers or completions. Select the one that is best in each case and then fill in the corresponding oval on the answer sheet.

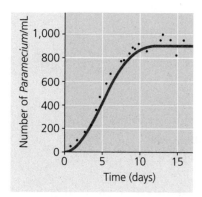

Paramecium Population in Lab

1. In the diagram above, what number of *Paramecium* best represents the carrying capacity of the environment for the population shown above?
(A) 200
(B) 500
(C) 600
(D) 900
(E) 1,000

2. If you assume that three genes—P, Q, and R—are not linked, that the probability of gene P appearing in a gamete is ¼, that the probability of gene Q appearing in a gamete is ¼, and that the probability of gene R occurring in a gamete is ¼, then which of the following is the probability that all three genes will appear in the same gamete?
(A) ¼ × ¼ × ¼
(B) ¼ + ¼ + ¼
(C) ¼ ÷ ¼ × ¼
(D) (¼)$^{1/3}$
(E) 0

3. A woman plans to have four children, and she hopes to have two girls first and then two boys. What is the probability that she will bear this ratio of offspring, and in this order?
(A) ½
(B) ¼
(C) $\frac{1}{16}$
(D) $\frac{1}{64}$
(E) $\frac{1}{128}$

GO ON TO THE
NEXT PAGE

4. All of the following are true about CO_2 in the atmosphere surrounding Earth EXCEPT
   (A) the process of photosynthesis in plants takes in CO_2 from the atmosphere
   (B) this CO_2 absorbs and re-reflects infrared rays from the sun
   (C) the level of CO_2 in the atmosphere is increasing, which is causing the earth's temperature to rise
   (D) the higher the level of CO_2 in the atmosphere, the faster the rate of growth of plants
   (E) the increase in CO_2 in the atmosphere is mostly due to the decrease in the number of plants on the earth's surface due to deforestation

5. Which of the following organelles is the site of cell respiration?
   (A) Golgi apparatus
   (B) Chloroplast
   (C) Mitochondria
   (D) Endoplasmic reticulum
   (E) Ribosomes

6. Which of the following describes how goslings recognize their mothers after they are born?
   (A) Habituation
   (B) Imprinting
   (C) Reasoning
   (D) Instinct
   (E) Maturation

7. A student using a light microscope observes a relatively small, rod-shaped cell that has genetic material not enclosed by a membrane. What type of cell is this most likely to be?
   (A) Viral
   (B) Eukaryote
   (C) Gamete
   (D) Prokaryote
   (E) Plant

8. The condition in which two species live in the same area but rarely encounter each other and do not interbreed is referred to as
   (A) latent variation
   (B) sterility
   (C) structural differences
   (D) habitat isolation
   (E) reproductive isolation

9. In a test cross, which of the following is true?
   (A) One of the individuals is homozygous dominant.
   (B) One of the individuals is homozygous recessive.
   (C) Both individuals are heterozygous.
   (D) Both individuals are homozygous.
   (E) Both individuals have an unknown phenotype.

10. Diatoms are the major primary producers in which of the following ecosystems?
    (A) Marine
    (B) Desert
    (C) Deciduous forest
    (D) Chaparral
    (E) Tropical rain forest

11. In deer, fur length is controlled by a single gene with two alleles. When a homozygous deer with long fur is crossed with a homozygous deer with short fur, the offspring all have fur of medium length. If these offspring with medium fur are mated with one another, what percentage of their offspring will have long fur?
    (A) 100%
    (B) 75%
    (C) 50%
    (D) 25%
    (E) 0%

12. All of the following are forms in which animals can secrete wastes EXCEPT
    (A) ammonia
    (B) urea
    (C) uric acid
    (D) feces
    (E) nitrate

13. Which of the following statements best supports the idea that certain cell organelles are evolutionarily derived from symbiotic prokaryotes living in host cells?
    (A) The process of cell respiration in certain prokaryotes is similar to that occurring in mitochondria and chloroplasts.
    (B) Mitochondria and chloroplasts have DNA and proteins that are very similar to those in prokaryotes.
    (C) Mitochondria and prokaryotes have similar cell wall structures.
    (D) Like prokaryotes, mitochondria have a double membrane.
    (E) Mitochondria and chloroplasts have similar DNA and ribosomes to prokaryotes.

14. During the course of which type of reaction is energy consumed?
    (A) Hydrolysis
    (B) Catabolic
    (C) Oxidation-reduction
    (D) Endergonic
    (E) Exergonic

15. Which of the following occurs during meiosis II but not during mitosis?
    (A) The diploid number of chromosomes is reduced to the haploid number.
    (B) The nuclear envelope disintegrates.
    (C) The chromatids of each chromosome are pulled apart.
    (D) Synapsis and crossing over take place.
    (E) Cytokinesis pinches off the cell membrane to produce two cells.

16. Consumption of CO_2 can be used as a measure of photosynthetic rate because carbon dioxide is
    (A) consumed during the light reactions of photosynthesis
    (B) consumed during the dark reactions of photosynthesis
    (C) used to trap photons, the form of energy in sunlight
    (D) necessary for the production of ATP in oxidative phosphorylation
    (E) necessary in order for fermentation to take place

17. Which of the following cell organelles is not bound by a membrane?
    (A) Centrosome
    (B) Golgi apparatus
    (C) Cell nucleus
    (D) Mitochondria
    (E) Peroxisome

18. Which type of leaf cell is the site of ATP synthesis?
    (A) Chlorophyll
    (B) Parenchyma
    (C) Guard cells
    (D) Mesophyll
    (E) Fusiform

19. Which of the following pairs of groups characterizes the structure of a fat?
    (A) Glycerin and three fatty acids
    (B) Glycerol and three fatty acids
    (C) Glycerin and two fatty acids
    (D) Glycerol and two fatty acids
    (E) Glycerol and one fatty acid

GO ON TO THE NEXT PAGE

20. Which of the following statements is NOT part of Darwin's theory of natural selection?
    (A) Individuals survive and reproduce with varying degrees of success.
    (B) Because there are more individuals than the environment can support, this leads to a struggle for existence in which only part of the offspring survives in each generation.
    (C) The unequal ability of individuals to survive and reproduce leads to a gradual change in the population.
    (D) Members of the population that are physically weaker than others will be eliminated first by forces in the environment.
    (E) Individuals in a population vary in their characteristics, and no two individuals are exactly alike.

21. In animal development, all of the following occur EXCEPT
    (A) cleavage, a succession of rapid cell divisions, occurs just after fertilization
    (B) the zygote develops polarity
    (C) the zygote develops into a hollow ball of cells called a blastula
    (D) cleavage continues, and produces a solid ball of cells called a morula
    (E) all of the genes in zygotic cells are activated

22. Which of the following describes a protein capable of converting related proteins to its infectious form?
    (A) Virus
    (B) Retrovirus
    (C) Prion
    (D) Spirochete
    (E) Prokaryote

23. Prokaryotic cells and eukaryotic cells differ in which of the following ways?
    (A) Membrane-bound nucleus
    (B) Genetic material in the form of DNA
    (C) Cytoplasm
    (D) Ribosomes
    (E) A cell membrane

24. Which of the following plant hormones is responsible for root growth and differentiation, cell division, germination, and delaying senescence?
    (A) Auxin
    (B) Cytokinins
    (C) Gibberellins
    (D) Abscisic acid
    (E) Ethylene

25. A farmer takes one green pepper plant that has all of the most desirable traits of the species, and the farmer produces a group of plants through asexual reproduction, using only genetic material from this ideal parent plant. The resulting group of plants are genetically identical to the parent and are said to be a
    (A) community
    (B) population
    (C) clone
    (D) phylum
    (E) genus

26. Archaebacteria, or Archae, contain prokaryotic organisms that
    (A) possess a nuclear envelope
    (B) have plantlike features
    (C) are capable of nitrogen fixation
    (D) live in extreme heat or acid environments
    (E) reproduce sexually

27. Oxytocin is a hormone that does which of the following in humans?
    (A) Stimulates growth
    (B) Stimulates the production and secretion of milk
    (C) Raises blood glucose levels
    (D) Lowers blood glucose levels
    (E) Stimulates contraction of the uterus

28. In dogs, the trait for long tail is dominant (L), and the trait for short tail is recessive (l). The trait for yellow coat is dominant (Y), and the trait for white coat is recessive (y). Mating two dogs gives a litter of 3 long-tailed, yellow dogs and 1 long-tailed white dog. Which of the following is most likely to be the genotype of the parent dogs?
    (A) $LLYY \times LLYY$
    (B) $LLyy \times LLYy$
    (C) $LlYy \times LlYy$
    (D) $LlYy \times LLYy$
    (E) $LlYY \times Llyy$

29. A light microscope can be used to view which of the following cell structures?
    (A) Ribosomes
    (B) Golgi apparatus
    (C) Nucleus
    (D) Lipids
    (E) Proteins

30. Which of the following is located at the tips of plant roots and shoots, and supplies cells through cell division for plant growth?
    (A) Lateral meristems
    (B) Apical meristems
    (C) Nodular meristems
    (D) Vascular bundles
    (E) Pericycle

31. Which of the following is a component of the plasma membrane of a cell?
    (A) Glycoproteins
    (B) Cytochromes
    (C) Nucleic acids
    (D) Phosphatidic acid
    (E) Lipoproteins

32. Mitosis in vertebrate cells occurs just after which of the following phases of the cell cycle?
    (A) G_1
    (B) S
    (C) DNA synthesis
    (D) G_2
    (E) M phase

33. The sporophyte generation is the dominant generation in which of the following?
    (A) Ferns
    (B) Mosses
    (C) Lichens
    (D) Liverworts
    (E) Hornworts

34. Certain cells of all of the following organisms undergo meiosis EXCEPT
    (A) ferns
    (B) sponges
    (C) fungi
    (D) bacteria
    (E) nematodes

35. Water and minerals flow up through a plant through the
    (A) sieve tubes of phloem
    (B) sieve tubes of xylem
    (C) tracheids and vessel elements of phloem
    (D) tracheids and vessel elements of xylem
    (E) vessel elements of xylem only

GO ON TO THE NEXT PAGE

36. In plants, changes in the level of which of the following causes the stomata to close and conserve water during drought?
    (A) Brassinosteroids
    (B) Abscisic acid
    (C) Auxin
    (D) Cytokinins
    (E) Ethylene

37. O_2 and CO_2 diffuse from regions where their partial pressures
    (A) are higher to regions where they are lower
    (B) are lower to regions where they are higher
    (C) are zero to regions of higher partial pressure
    (D) are zero to regions of lower partial pressure
    (E) are influenced by external atmosphere changes into the cell

38. A DNA molecule that can carry foreign DNA into a cell and then replicate is called a
    (A) probe
    (B) restriction fragment
    (C) restriction enzyme
    (D) vector
    (E) transcriptase

39. It is theorized that which of the following was relatively sparse in the Earth's atmosphere when molecules that could replicate themselves originated?
    (A) Gaseous oxygen
    (B) Water
    (C) Carbon dioxide
    (D) Nitrogen
    (E) Hydrogen

40. The composition of lymph, in lymph vessels, is roughly the same as which of the following?
    (A) Blood
    (B) Interstitial fluid
    (C) Glomerular filtrate
    (D) Bile
    (E) Chyme

41. In photosynthesis, most ATP is produced as a result of which of the following processes?
    (A) The light reactions
    (B) Carbon fixation
    (C) Noncyclic photophosphorylation
    (D) The dark reactions
    (E) The Calvin cycle

42. Which of the following organisms is not usually considered alive because of its dependence on other organisms for reproduction?
    (A) *E. coli*
    (B) Tapeworm
    (C) Mold
    (D) Virus
    (E) Lichen

43. In guinea pigs, black fur (*B*) is dominant to brown fur (*b*). No tail (*T*) is dominant over tail (*t*). What fraction of the progeny of the cross *BbTt* x *BbTt* will have black fur and tails?
    (A) $\frac{1}{16}$
    (B) $\frac{3}{16}$
    (C) $\frac{3}{8}$
    (D) $\frac{9}{16}$
    (E) 1

44. While migrating, a flock of Canadian geese that constitute the entire population of a certain geographic area in New Hampshire is nearly decimated by a tornado that they encounter. Only four of the geese survive, and they continue their migration. When they return to their territory in New Hampshire in the spring, these four start a new colony. This phenomenon is known as
    (A) the founder effect
    (B) natural selection
    (C) migration
    (D) polymorphism
    (E) the bottleneck effect

45. Which type of plant is more likely to grow under hot, arid conditions?
    (A) Bryophyte
    (B) C$_4$ plant
    (C) A plant with a thin, light epidermal layer
    (D) C$_3$ plant
    (E) Angiosperm

46. Which of the following describes a drawing showing the evolutionary history among a particular species or a group of related species?
    (A) Food web
    (B) Food chain
    (C) Pedigree
    (D) Phylogeny
    (E) Graph

47. Which of the following characteristics is common to all bryophytes?
    (A) Large, independent gametophytes
    (B) Large, independent sporophytes
    (C) Haploid spores
    (D) Seed production
    (E) Fertilization in water

48. The rate of flow of sugar and nutrients through the phloem is regulated by
    (A) diffusion from source to sink
    (B) hydrostatic pressure in the sieve tube
    (C) the force of transpirational pull
    (D) active transport by tracheid and vessel cells
    (E) passive transport by the pith

49. It is theorized that which of the following metabolic pathways evolved first?
    (A) The Krebs cycle
    (B) The Calvin cycle
    (C) Glycolysis
    (D) Oxidative phosphorylation
    (E) The electron transport chain

50. Allolactose stimulates the cells of the human body to produce mRNAs that code for the enzyme B-galactosidase, which breaks down lactose into glucose and galactose. In this case, the role of allolactose can best be described as that of a
    (A) DNA replication stimulator
    (B) translation inhibitor
    (C) stimulator of B-galactosidase secretion
    (D) regulator of gene activity
    (E) translation activator

51. Which of the following types of data can be used to map the locations of genes on chromosomes?
    (A) Segregation frequency
    (B) Rate of gene regulation
    (C) Dominance patterns
    (D) Rate of gene recombination
    (E) Rate of gene expression

52. When the stomata of a plant leaf open, which of the following occurs?
    (A) There is a decrease in CO$_2$ intake in the leaf.
    (B) The plant shifts from C$_3$ photosynthesis to C$_4$ photosynthesis.
    (C) The rate of transpiration decreases.
    (D) There is an increase in the concentration of CO$_2$ in mesophyll cells.
    (E) There is an increase in the rate of production of nucleic acids.

53. All of the following are evidence for evolution EXCEPT
    (A) the presence of anatomical homologies
    (B) the existence of embryological homologies
    (C) the existence of molecular homologies
    (D) the fossil record
    (E) the existence of homologies in diet among species

GO ON TO THE NEXT PAGE

54. In a certain group of iguanas, the presence of brown skin is the result of a homozygous recessive condition in the biochemical pathway producing hair pigment. If the frequency of the allele for this condition is 0.35, which of the following is closest to the frequency of the dominant allele in this population? (Assume that the population is in Hardy-Weinberg equilibrium.)
    (A) 0.15
    (B) 0.45
    (C) 0.55
    (D) 0.65
    (E) 0.85

55. A biologist isolates many small, oblong, single-celled photoautotrophic protists from a murky pond. These cells have flagella. Looking at these organisms under an electron microscope will reveal the presence of
    (A) a central vacuole and chloroplasts
    (B) ribosomes and chloroplasts
    (C) ribosomes and mitochondria
    (D) chloroplasts and mitochondria
    (E) lysosomes and mitochondria

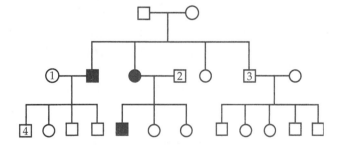

56. In the pedigree above, squares represent males, and circles represent females. Shaded figures represent individuals who possess a particular trait. Which of the following patterns of inheritance best explains how this trait is transmitted?
    (A) Partially dominant
    (B) Autosomal dominant
    (C) Autosomal recessive
    (D) Sex-linked recessive
    (E) Sex-linked dominant

57. The movement of H^+ ions across the inner mitochondrial membrane, from the mitochondrial membrane to the intermembrane space during chemiosmosis of cell respiration, is an example of what type of movement across a membrane?
    (A) Active transport
    (B) Facilitated diffusion
    (C) The work of a symport
    (D) The work of an antiport
    (E) Cotransport

58. Which of the following is the most direct result of the presence of protein in the small intestine?
    (A) The secretion of bile by the gallbladder
    (B) The secretion of pepsin by the lining of the small intestine
    (C) The activation of the inactive form of trypsin and chymotrypsin to their active form
    (D) The activation of the inactive form of lipase to its active form
    (E) Peristalsis along the walls of the small intestine

59. A particular tapeworm has a life cycle in which the adult lives in a human's intestine, attaching itself to the lining of the digestive tract, absorbing nutrients that are digested by the host, and then releasing eggs that are secreted in the human's feces. This feces contaminates the food of a pig, and the larvae encysts in the muscle of the pig, which is later consumed by humans. The tapeworm is an example of a
    (A) mutualistic symbiotic partner to humans
    (B) commensalistic symbiotic partner to humans
    (C) parasitic symbiotic partner to humans
    (D) mutualistic symbiotic partner to pigs
    (E) commensalistic symbiotic partner to pigs

60. Organisms that can live in the benthic zone in the ocean primarily consume which of the following?
    (A) Algae
    (B) Phytoplankton
    (C) Zooplankton
    (D) Detritus
    (E) Cyanobacteria

61. Which of the following cellular processes is coupled with active transport?
    (A) The addition of H^+ to H_2O to produce hydronium ion
    (B) The dephosphorylation of ATP
    (C) The phosphorylation of ADP
    (D) The dephosphorylation of G3P
    (E) The formation of peptide bonds between amino acids

62. Which of the following is thought to be human beings' closest evolutionary relative?
    (A) Chimpanzee
    (B) Gorilla
    (C) Bonobo
    (D) Orangutan
    (E) Gibbon

63. Which of the following cells would most likely have the greatest concentration of mitochondria in its cytoplasm?
    (A) A cell lining the digestive tract
    (B) An active skeletal muscle cell
    (C) A cell in the liver
    (D) A cell in the brain
    (E) A cell in the epidermis

64. In which of the following pairs are the organisms most closely related taxonomically?
    (A) Mushroom; tulip
    (B) *E. coli*; Euglena
    (C) Lobster; spider
    (D) Shark; crayfish
    (E) Dolphin; sea star

GO ON TO THE
NEXT PAGE

Directions: Each group of questions below consists of five lettered headings (or five lettered items in a graph) followed by a list of numbered phrases or sentences. For each numbered phrase or sentence select the one heading (or item) that is most closely related to it and fill in the corresponding oval on the answer sheet. Each heading may be used once, more than once, or not at all in each group.

Questions 65–68 refer to the following graph. Each of the curves represents one pathway for the same reaction, but one pathway is catalyzed by an enzyme.

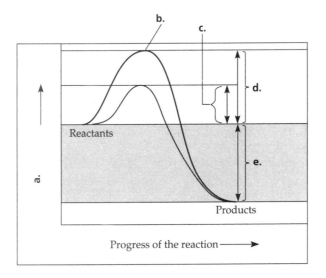

65. Represents the activation energy of the uncatalyzed reaction

66. Represents the activation energy of the catalyzed reaction

67. Represents the free energy of the participants in the reaction

68. Represents the transition state of the uncatalyzed reaction

Questions 69–73 refer to the following diagram of the structure of a flower.

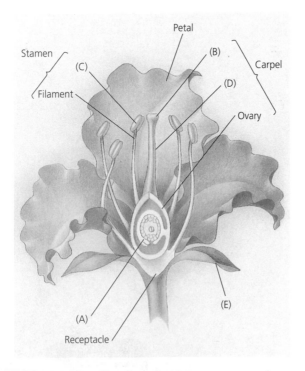

Structure of a Flower

69. Develop into seeds after fertilization

70. The site of pollen production

71. Receives pollen

72. Enclose the flower prior to its opening

73. Leads to the ovary

Questions 74–77

    (A) Follicle-stimulating hormone (FSH)
    (B) Growth hormone (GH)
    (C) Melatonin
    (D) Androgens
    (E) Endorphins

74. Secreted by the anterior pituitary gland, stimulates growth and metabolism

75. Secreted by the anterior pituitary gland, stimulates the production of ova and sperm

76. Secreted by the testes, promotes the development of secondary sex characteristics

77. Secreted by the pineal gland, involved in regulating biological rhythms

Questions 78–81

    (A) Savannah
    (B) Taiga
    (C) Tundra
    (D) Chaparral
    (E) Coniferous forest

78. Dominated by short shrubs and grasses, endures long dark winters

79. Home to plant life adapted to periodic fires and to dense evergreen shrubs

80. Has long, cold winters and short summers, a biome that is dominated by gymnosperms

81. Home to large herbivores and their predators, marked by grasses and scattered trees.

Questions 82–86

    (A) Inner mitochondrial membrane
    (B) The cytoplasm
    (C) Thylakoid membranes
    (D) Ribosome
    (E) Nucleus

82. Where mRNA is translated into proteins

83. Where DNA is replicated prior to cell division

84. The location of the electron transport chain

85. Where photophosphorylation takes place

86. The location of glycolysis

Questions 87–91

    (A) Rotifera
    (B) Porifera
    (C) Nematoda
    (D) Platyhelmintha
    (E) Chordata

87. Possess a notochord, a dorsal hollow nerve cord, and bilateral symmetry

88. Possess a tough exoskeleton called a cuticle, are not segmented, have a complete digestive tract but no circulatory system

89. Possess a complete digestive tract, are pseudo-coelomates with a crown of cilia surrounding their mouths

90. Are hermaphrodites and suspension-feeders, have no nerves or muscles, have a sac-like body

91. Possess a gastrovascular cavity with only one opening, are acoelomates, and include many parasitic species

GO ON TO THE
NEXT PAGE

Questions 92–95
    (A) Telomere
    (B) DNA polymerase
    (C) Helicase
    (D) Primer
    (E) DNA ligase

92. DNA made up not of genes, but of multiple repetitions of short nucleotide sequences

93. Joins the sugar-phosphate backbones of the Okazaki fragments to create a complete DNA strand

94. Catalyzes elongation of new DNA at a replication fork

95. Catalyzes the unwinding of double-stranded DNA prior to transcription

Questions 96–99
A scientist is studying the cell cycles of various organisms to learn about their metabolic activities and division patterns. She kept track of the amount of time each type of cell spent in the cell cycles and collected them in the table below.

TOTAL MINUTES SPENT IN EACH CELL CYCLE PHASE				
Cell Type	G_1	S	G_2	M
Monkey liver	20	23	10	18
Plant stem	98	0	0	0
E. coli	18	36	14	19

96. From the data in the table above, which of the following is the most likely conclusion about the cell from the plant stem?
    (A) It is dead.
    (B) This cell contains no DNA.
    (C) This cell contains no mRNA.
    (D) This cell has entered the G_0 phase.
    (E) This cell is continually growing.

97. How long did the entire process of mitosis take in the monkey liver cell?
    (A) 20 minutes
    (B) 23 minutes
    (C) 18 minutes
    (D) 10 minutes
    (E) 0 minutes

98. How long does it take for the E. coli cell to copy all of its chromosomes?
    (A) 18 minutes
    (B) 36 minutes
    (C) 14 minutes
    (D) 19 minutes
    (E) 0 minutes

99. What does this data tell you about E. coli, compared to the genome of the monkey?
    (A) The E. coli genome is much larger than the monkey genome.
    (B) The monkey genome is much larger than the E. coli genome.
    (C) The two genomes are of comparable size.
    (D) DNA replication occurs much more quickly in E. coli than in monkey liver cells.
    (E) DNA replication occurs much more quickly in monkey liver cells than in E. coli.

Questions 100–101
In a study of the development of chicken embryos, groups of cells in the early germ layers were stained with five different-colored dyes. After the organs of the chick developed, the location of the dyes were marked down as shown below.

Tissue	Color
Brain	Blue
Liver	Red
Mucous membranes	Green
Nerve cord	Yellow
Heart	Orange

100. Mesoderm would eventually give rise to
    tissues containing which of the following
    colors?
    (A) Yellow and purple
    (B) Red and blue
    (C) Orange and yellow
    (D) Orange and green
    (E) Red and green

101. Tissues that were stained blue were derived
    from
    (A) mesoderm
    (B) ectoderm
    (C) mesoderm and ectoderm
    (D) ectoderm and endoderm
    (E) mesoderm and endoderm

*Questions 102–103 refer to the following chromo-
some map.*

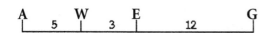

102. Considering the possibilities of recombina-
    tion, which two genes on this chromosome
    are most likely to segregate together into a
    daughter cell?
    (A) A and W
    (B) A and E
    (C) A and G
    (D) W and E
    (E) W and G

103. If the rate of recombination between gene A
    and W is 5%, what is the rate of recombina-
    tion between gene W and gene G?
    (A) 0%
    (B) 5%
    (C) 10%
    (D) 15%
    (E) 20%

*Questions 104–106 refer to the following figure,
which shows a food web in a particular ecosystem.
Each letter represents a species in this ecosystem, and
the arrows show the flow of energy.*

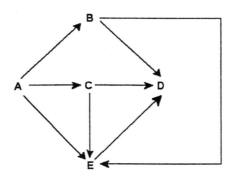

104. Which of the species in the above food web is
    the primary producer?
    (A) A
    (B) B
    (C) C
    (D) D
    (E) E

105. Species B and C represent which of the
    following?
    (A) Primary producers
    (B) Primary consumers
    (C) Secondary consumers
    (D) Tertiary consumers
    (E) Omnivores

106. Which of the following most accurately
    describes species E?
    (A) An herbivore, and a secondary consumer
    (B) An omnivore, and a secondary consumer
    (C) An omnivore, and both a primary and
        secondary consumer
    (D) An herbivore, and both a primary and
        secondary consumer
    (E) A cannibal

Questions 107–110 refer to the information about five organisms shown in the table below.

Environment Inhabited by Animal	Body Length	Features of Gas Exchange System	Features of Gas Exchange Surface	Percentage of Oxygen Extracted from Air
1 Terrestrial	0.01 m	Branching air tubes, large tracheae that open to the outside	Moist epithelium lining the terminal ends of the tracheal system	53%
2 Terrestrial	0.02 m	Branching air tubes, large tracheae, ventilates with rhythmic body movements	Moist epithelium lining the terminal ends of the tracheal system	48%
3 Aquatic	0.5 m	Outfoldings in the body surface suspended in water	Uses countercurrent exchange and ventilation	73%
4 Terrestrial	1.0 m	Lungs that work in conjunction with circulatory system	Gas exchange occurs across epithelium of alveoli	67%
5 Terrestrial	2.0 m	Lungs that work in conjunction with circulatory system	Gas exchange occurs across pithelium of alveoli	78%

107. Which of the following can be concluded based on the data from the table above?
(A) Lungs are more efficient than trachea at extracting oxygen from the air.
(B) Lungs are less efficient than trachea at extracting oxygen from the air.
(C) Gills are more efficient than lungs at extracting oxygen from the air.
(D) Gills are more efficient than lungs at extracting oxygen from the air.

108. Which of the above organisms is most likely to have hemolymph as its main circulatory fluid?
(A) 1 or 2
(B) 2 or 3
(C) 3 or 4
(D) 4 or 5
(E) None of these organisms

109. In which of these organisms is hemoglobin used to transport oxygen through the blood?
(A) 5 only
(B) 4 and 5 only
(C) 3, 4, and 5 only
(D) 2, 3, 4, and 5 only
(E) All of them

110. In which of these animals can the process of gas exchange occur without physical movement of some part of the animal?
(A) 1 and 2 only
(B) 1 and 3 only
(C) 1, 2, and 3 only
(D) 3, 4, and 5 only
(E) All of them

Questions 111–113 refer to the following gel, which was produced from four samples of radioactively-labeled DNA that were cut with one type of restriction enzyme. The samples were separated by gel electrophoresis. Answer the questions on the basis of the bands you can visualize below.

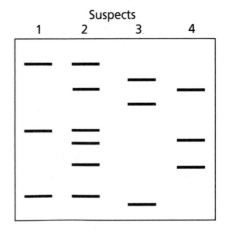

111. The DNA fragments in the gel were separated when an electric field was applied across the gel and they migrated at different speeds. The differential migration speed of the different DNA fragments was due to the
    (A) amount of radioactivity in the samples
    (B) degree to which the samples were negatively charged
    (C) degree to which the samples were positively charged
    (D) size of the samples
    (E) polarity of the samples

112. Which of the following is true about the DNA samples that were loaded onto the gel?
    (A) The DNA strand of sample 2 was originally the longest.
    (B) The DNA strand of sample 4 was originally the shortest.
    (C) Samples 2 and 4 are the same DNA sample.
    (D) Sample 2 was cut at more restriction sites than was sample 4.
    (E) Sample 4 was cut at more restriction sites than was sample 2.

113. What was the purpose in radioactively labeling the DNA fragments in this experiment?
    (A) To visualize them
    (B) To make them travel through the gel
    (C) To hydrolyze them into fragments
    (D) To get rid of contaminants
    (E) To destroy their polarity

Questions 114–116 refer to an experiment in which there is an initial setup of a U-tube with its two sides separated by a membrane that permits the passage of water and small particles (such as NaCl) but not of large molecules (such as glucose). The U-tube is filled on one side with a solution of 0.4M glucose and 0.5M NaCl, and on the other, 0.8M glucose and 0.4M NaCl.

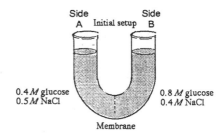

114. When this U-tube was set up, at time = 0 in the experiment, which of the following was definitely true?
    (A) The solution on side A was more concentrated than the solution on side B.
    (B) The solution on side B was more concentrated than the solution on side A.
    (C) The two solutions had equal concentration.
    (D) There was more salt on side B than on side A.
    (E) There was more glucose on side A than on side B.

115. Which of the following is most likely to occur after two hours of the U-tube being undisturbed?
    (A) The water levels of sides A and B will remain the same.
    (B) The amount of NaCl on side B will have increased.
    (C) The amount of NaCl on side A will have increased.
    (D) The amount of glucose on side B will have increased.
    (E) The amount of glucose on side A will have increased.

GO ON TO THE
NEXT PAGE

116. After two hours, which of the following would probably be true of the level of water on each side of the U-tube?
    (A) There would be no change in the water levels on either side of the U-tube.
    (B) The water column in side A would be slightly higher.
    (C) The water column in side B would be slightly higher.
    (D) The water columns on both sides would be slightly lower.
    (E) The water columns on both sides would be slightly higher.

Questions 117–120 refer to the graph below, which shows the change in two different populations of wild boars that were introduced to an isolated geographic area in 1905.

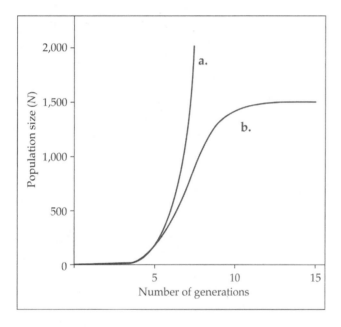

117. The type of population growth exhibited by population A is
    (A) logistic
    (B) stable
    (C) density-dependent
    (D) exponential
    (E) diversifying

118. Population B reaches its carrying capacity at about what size?
    (A) 0 members
    (B) 500 members
    (C) 1,000 members
    (D) 1,500 members
    (E) 2,000 members

119. The drop in numbers of population B at carrying capacity is due most likely to which of the following?
    (A) Density-dependent factors
    (B) Density-independent factors
    (C) Massive viral infection in the population
    (D) Natural disaster
    (E) Migration

120. The split that occurred in the two populations may have been the direct result of
    (A) convergent evolution
    (B) speciation
    (C) segregation
    (D) geographic isolation
    (E) bottleneck effect

END OF SECTION I

Biology
Section II

Time—1 hour and 30 minutes

Answer all questions. Number your answer as the question is numbered below.

Answers must be in essay form. Outline form is NOT acceptable. Labeled diagrams may be used to supplement discussion, but in no case will a diagram alone suffice. It is important that you read each question completely before you begin to write.

1. Birth control pills are chemical contraceptives that are made up of estrogen and progestin (which is a progesterone-like substance). They act through a negative feedback loop to stop the secretion of GnRH by the hypothalamus, and of FSH and LH by the pituitary.
   (a) Explain how a negative feedback system works.
   (b) Explain how the effects of the birth control pill described above would make pregnancy highly unlikely.

2. Gene expression in a cell is influenced by a variety of factors—not all genes on the eukaryotic chromosome are expressed, and in fact only a small fraction of the genes are transcribed into working proteins.
   (a) Discuss three ways in which gene control works in the cell.
   (b) Describe three lab procedures you could employ in order to determine how much transcription and translation is going on in a cell at a given time.

3. It has been determined that, evolutionarily, the closest relative of humans is the chimpanzee. Other somewhat close relatives are the gorilla and the orangutan.
   (a) Describe the relationships among these four species—taxonomically and through phylogeny.
   (b) Describe three kinds of evidence that were used to determine the relationship among these four species.
   (c) Describe the general structure of the ancestor of *Homo sapiens*, relative to that of other arthropods.

4. A flowering plant in a ceramic pot is placed in a window that faces east, and it is given adequate water and soil nutrients.
   (a) Describe the daily and nightly events in the plant's metabolism.
   (b) Describe the changes in the plant that would be induced by moving the plant to a window that faced west.

END OF EXAMINATION

ANSWERS AND EXPLANATIONS

Multiple-Choice Questions

1. (D) is correct. The carrying capacity of a population is defined as the maximum population size that a certain environment can support without itself being degraded. If you look at the graph, you can see that the population increases in number until it reaches about 900 members, and at that point it stabilizes. The number 900 therefore represents the carrying capacity.

2. (A) is correct. This is a simple probability question. In order to calculate the chance that two or more independent events will occur together in a specific combination, you can use the multiplication rule. Take the probability that gene P will segregate into a gamete ($\frac{1}{4}$), and multiply it by the probability that gene Q will segregate into a gamete ($\frac{1}{4}$). Then multiply that by the probability that gene R will segregate into a gamete to get $\frac{1}{4} \times \frac{1}{4} \times \frac{1}{4} = \frac{1}{64}$.

3. (C) is correct. The probability of this woman bearing two daughters and then two sons is 1/16. You can get this answer by using the rule of multiplication. Whether the woman will bear a boy or girl each time she bears a child will *not* be influenced by whether the child before or after the one in question was a girl or boy—in other words, each pregnancy and birth is an independent event. Therefore, each time the woman bears a child, there will be a $\frac{1}{2}$ probability that it will be the sex she wants it to be. Therefore, multiply $\frac{1}{2} \times \frac{1}{2} \times \frac{1}{2} \times \frac{1}{2} = \frac{1}{16}$.

4. (E) is correct. All of the answer choices are true except choice *E*. It is theorized that the increase in CO_2 in the Earth's atmosphere is due to the fact that human beings have been burning fossil fuels since the Industrial Revolution. This increase in CO_2 in the Earth's atmosphere is leading to a slight but gradual increase in the global temperature, causing global warming.

5. (C) is correct. The mitochondria of the cell is the site of cell respiration, and it is the organelle that produces the most ATP in the cell by extracting energy from sugars, fats, and other fuels. Mitochondria are found in almost all eukaryotic cells, and they are enclosed by two membranes—a smooth outer membrane and an inner membrane with many infoldings called cristae. The inside of the inner membrane is called the mitochondrial matrix. There are two processes that occur in the mitochondrion to produce energy. One is the Krebs cycle, and the other is chemiosmosis.

6. (B) is correct. Imprinting is defined as a type of learning that is generally irreversible and that is limited to a certain period in an animal's life (usually when the animal is very young). The phenomenon of mother-offspring bonding in geese is an example of this type of learning. If this imprinting does not happen, the mother will not take care of the offspring, and the goslings will die.

7. (D) is correct. This is most likely a prokaryotic cell. Most prokaryotes are about one-tenth the size of eukaryotic cells, and the most common shapes of prokaryotic cells are spherical, rod-shaped, and helical. Finally, prokaryotes lack membrane-bound organelles, including nuclear membranes, to contain their genetic material; instead they have nucleoid regions, which are a complex of DNA and fibers in a certain region of the cell.

8. (D) is correct. Habitat isolation is an example of a prezygotic barrier to mating between two species, in which the two species live in the same geographic area, but rarely encounter each other, and thus do not interbreed. These two species are not geographically isolated, but one might be aquatic while the other is terrestrial, for example.

9. (B) is correct. A testcross is the breeding of an organism that has an unknown genotype to one that is homozygous recessive, in order to determine the genotype of the unknown parent. The phenotypic ratio of the offspring will reveal the unknown genotype.

10. (A) is correct. Diatoms are single-celled chrysophytes that reproduce asexually by mitosis. They live in both freshwater and marine environments, and they are very numerous in those environments. They are the major primary producers in marine environments mostly because of the fact that they are the most numerous.

11. (D) is correct. If you consider that the homozygous long hair deer is *HH*, and the homozygous short hair deer is *hh*, then crossing them would give all offspring with the genotype *Hh*, and this means that the heterozygotes have medium-length hair. Crossing the heterozygotes would give you offspring in the ratio of 1:2:1—*HH:Hh:hh*. This means that 25% of the offspring would have long hair (*HH*), 25% of them would have short hair (*hh*), and 50% of them would have medium-length hair (*Hh*).

12. (E) is correct. Nitrate is the only answer listed that does not represent a form of waste secreted by some kind of animal. Ammonia is the waste product secreted by many aquatic species; urea is the common waste form of mammals, most amphibians, and many fishes; and birds and reptiles secrete uric acid. Solid waste in the form of feces is also excreted by many animals.

13. (E) is correct. Among these answers, the one that best supports the idea that certain cell organelles, like mitochondria and chloroplasts, were once symbiotic prokaryotes living inside larger cells is answer *E*, which states that mitochondria and chloroplasts have similar DNA and chromosomes to bacteria.

14. (D) is correct. An endergonic reaction is a nonspontaneous chemical reaction; in order for the reaction to go, free energy must be absorbed from the surroundings. An exergonic reaction is a spontaneous reaction; during the course of the reaction energy is given off.

15. (A) is correct. During meiosis II, the diploid number of chromosomes is halved, and haploid cells, or gametes, are produced. Later these gametes can fuse during fertilization to form a diploid zygote. During mitosis, the parent cell is diploid, and after DNA replication one round of cell division occurs to produce two diploid daughter cells.

16. (B) is correct. The Calvin cycle—often referred to as the dark reactions of photosynthesis—occurs in the stroma of the chloroplast. In the course of the cycle, the enzyme rubisco combines carbon dioxide with a five-carbon sugar, consuming NADPH and ATP, and ultimately producing glyceraldehyde-3-phosphate. Since carbon dioxide is consumed in the course of the dark reactions of photosynthesis, the rate of its consumption can be used to determine the rate of photosynthesis.

17. (A) is correct. All of the cell organelles listed are contained by a membrane composed of phospholipids except centrosomes. Centrosomes are areas that are present in the cell cytoplasm only during cell division; they are the sites at which the centrioles containing microtubules comprising the spindle apparatus are organized.

18. (B) is correct. The site of ATP synthesis in the leaf is the parenchyma cell. Parenchyma cells contain many chloroplasts, which are the plant cell organelles associated with and especially equipped for photosynthesis. There are two types of parenchyma cells, organized according to their location in the leaf—spongy parenchyma and palisade parenchyma.

19. (B) is correct. Fats, also known as triglycerides, are made up of three hydrophobic fatty acid molecules attached to a hydrophilic glycerol head group. Saturated fats contain no double bonds in their fatty acid chains, whereas unsaturated fats have one or more double bonds in their fatty acid chains. Phospholipids, which make up membranes, are similar to fats but have two fatty acid chains and a phosphate group attached to a glycerol molecule.

20. (D) is correct. The only one of the statements not included in Darwin's theory of natural selection is answer *D*. Darwin's theory stated that the individuals that are least well suited will not survive to reproduce, but physical weakness does not necessarily lead to an individual's being unsuited for its environment.

21. (E) is correct. All of the answers list events in the embryonic development of an animal except *E*. During development, certain genes are turned on at certain points in the process, but at no point are all of the genes in a cell activated.

22. (C) is correct. Prions are proteins that are the cause of diseases such as "mad cow" disease; they are misfolded proteins that are capable of converting normally folded proteins into misfolded forms (like themselves), triggering a chain reaction that vastly increases their numbers. These proteins exist primarily in the brain, which is why their misfolding has such serious negative effects.

23. (A) is correct. Prokaryotic cells are very simple cells that have no membrane-bound nucleus—instead they have a nucleoid region, at which the genetic material is concentrated. They do, however, share the remaining characteristics with eukaryotes. That is, they are bound by a membrane; they have genetic material that is translated into proteins on ribosomes; and they have cytoplasm.

24. (B) is correct. Cytokinins enhance the growth and development of plant cells; they are produced primarily in actively growing tissues like roots, embryos, and the fruits of a plant. They act in concert with auxins to cause cell division and differentiation. They are also responsible for retarding the aging process of some plant organs.

25. (C) is correct. Asexual reproduction is a form of reproduction in which just one parent contributes genetic material to the offspring, and the individual that reproduces asexually will give rise to a clone, which is a group of genetically identical individuals. On the other hand sexual reproduction, in which two parents contribute genetic material to the offspring, results in offspring that differ from each other and their parents.

26. (D) is correct. The domain Archaea, the archaebacteria, possesses members that are known for their extreme sturdiness. Many Archae are extremophiles: they thrive in extreme environments such as hot-water geysers in Yellowstone Park. Extreme halophiles live in very saline places, extreme thermophiles live in very hot environments.

27. (E) is correct. Oxytocin is produced in the posterior pituitary gland and regulated by the nervous system. It stimulates the powerful contractions of the smooth muscles in the uterine wall that occur during childbirth. Oxytocin also stimulates the placenta to secrete prostoglandins, which also contribute to contractions.

28. (D) is correct. The ratio of offspring produced is 3:1. Answer *A* would produce all dogs with long tails and yellow coats, since only the dominant alleles are present. That can be eliminated. First figure out the gametes that the resulting parents in each answer would produce, and do a Punnett square to figure out the resulting offspring proportions. For answer *D*, the gametes produced by parent 1 are *LY, Ly, lY*, and *ly*. For parent two, the possible gametes are *LY, LY, LY*, and *Ly*.

29. (C) is correct. The light microscope can be used to view structures that are between about 100 mm and 1 mm in size. This means that they can be used to see most plant and animal cells, cell nuclei, and certain large organelles like mitochondria. They can also be used to view bacterial cells, which are much smaller than eukaryotic cells.

30. (B) is correct. Apical meristems are located at the roots and shoots of plants, and they supply cells that enable the plant to grow in length. The elongation of plants is called primary growth (as opposed to secondary growth, when plants grow in diameter). Meristems are perpetually embryonic tissues that exist in a plants areas of growth.

31. (A) is correct. Glycoproteins are complexes of carbohydrate and protein that are associated with the cell membrane and that function in cell-cell recognition—a cell's ability to determine the function of a neighboring cell—for example. Glycoproteins vary in nature from species to species and from one cell type of an individual organism to another.

32. (D) is correct. Mitosis, or the M phase of the cell cycle, occurs just after the G_2 phase in the cell cycle. In the G_1 phase, the cell grows; in the S phase, the cell continues to grow and also copies its chromosomes; and in the G_2 phase, again the cell grows, and it begins to prepare for cell division. It divides finally in the M phase.

33. (A) is correct. The sporophyte generation of a plant is the one in which the cells of the plant are diploid, meaning that they have two sets of chromosomes. Ferns are the only species listed in which the sporophyte generation is dominant. The fern plant (which you can see growing in the ground) is diploid, whereas the gametophyte is a small unnoticeable plant on the earth's floor.

34. (D) is correct. Bacteria, which are the only prokaryotes listed, reproduce by an asexual reproductive process called binary fission. In binary fission, the bacterial cell chromosome (which consists of a single circular DNA molecule) replicates itself, and the cell grows with the plasma membrane growing inward and eventually pinching off to form two cells. The organisms in all of the other answers produce gametes, and gametes are produced by meiosis.

■ **35. (D) is correct.** The water-conducting elements of xylem (which is the plant tissue that carries water upwards from the roots toward the shoots of the plant) are the tracheids and vessel elements. These cells are elongated and dead when they reach functional maturity. When the cell dies, its interior disintegrates. This leaves only the hard cell wall, which forms a conduit for the movement of water.

■ **36. (B) is correct.** Abscisic acid is a plant hormone that prompts seeds to enter the dormant phase, which allows them to wait for favorable conditions before germinating. Abscisic acid also acts as the primary hormonal signal in times of drought—when a plant begins to wilt, abscisic acid accumulates in leaves and causes the stomata to close, which prevents water loss through transpiration.

■ **37. (A) is correct.** Gases in general, including carbon dioxide and oxygen, diffuse down a pressure gradient from regions where their partial pressure is higher to regions where they are lower. This means that if cells in one location are depleting their supply of oxygen, the partial pressure of oxygen in that area will drop, and more oxygen will be unloaded at the site of the depletion.

■ **38. (D) is correct.** A vector is an important component of genetic engineering; it is a plasmid (a circular piece of DNA) that is capable of having a foreign piece of DNA inserted into it. The vector can then be injected into a bacterial cell, causing transformation, and can then be replicated (along with the DNA fragment that it carries) by the cell's machinery.

■ **39. (A) is correct.** It is theorized that diatomic oxygen levels in the atmosphere were quite low around the time when the first self-replicating organisms were appearing on the Earth. Oxygen began to accumulate in the atmosphere when photosynthesis evolved. Photosynthesis split water to produce diatomic oxygen, which was released into the atmosphere and began to accumulate.

■ **40. (B) is correct.** Lymph is the fluid that is inside the lymphatic system, and the composition of lymph is similar to that of interstitial fluid. Fluids and proteins that are lost as they pass through the capillaries are picked up by the lymph vessels and returned to the blood. The lymphatic system drains into the circulatory system near the junction of the venae cavae and the right atrium.

■ **41. (C) is correct.** In noncyclic photophosphorylation, ATP is generated by chemiosmosis; the redox reactions of the electron transport chain create an H^+ gradient across the thylakoid membrane, and this gradient is used to power an ATP synthase. This ATP synthase makes ATP.

■ **42. (D) is correct.** Viruses are usually not considered to be alive, because they depend on host cells in order to reproduce. They inject their DNA into the host cell's nucleus, and their genetic material is replicated along with that of the host cell. Viral replication follows either a lytic cycle or a lysogenic cycle, either destroying its host or leaving the host intact.

■ **43. (B) is correct.** Crossing guinea pigs with genotype *BbTt* and *BbTt* gives a ratio of offspring of 9:3:3:1; this is a dihybrid cross between two independently assorting characters. Nine-sixteenths of the offspring will display both of the dominant traits; 3/16 of the offspring will display one of the dominant traits and one recessive one; 3/16 of the offspring will display the other dominant trait and the other recessive trait; and 1/16 of the offspring will display both recessive traits. This question asks how many of the offspring will display one of the dominant traits (black fur) and one of the recessive traits (tail); the answer is 3/16.

■ **44. (E) is correct.** This is an example of the bottleneck effect—a type of genetic drift. Genetic drift is defined as a change in the allelic frequencies of a population due to chance, and the bottleneck effect occurs when a large part of a certain population is destroyed by a disaster such as an earthquake, drought, or fire. The surviving members may not be representative of the population's gene pool, and they reproduce to form the new population.

■ **45. (B) is correct.** C_4 plants have evolved to thrive in hot, arid climates. In C_4 plants, photosynthesis is prefaced by an alternate mode of carbon fixation, in which a four-carbon compound is formed. Another plant that thrives in hot climates is the CAM plant. CAM plants (such as cactus) take in carbon dioxide at night and incorporate it into organic acids.

■ **46. (D) is correct.** A phylogeny is the evolutionary history of a species or a group of related species. Systematists create phylogenies in order to study the path of evolution on earth and to better understand relationships among species that live on earth today.

■ **47. (A) is correct.** The gametophyte is the dominant generation in bryophytes and typically is a large, visualized plant. Bryophytes produce diploid spores that land on suitable environments and divide by mitosis to eventually grow into gametophytes. Most bryophytes lack vascular tissue, which limits their ability to grow tall. Some common bryophytes are mosses, liverworts, and hornworts.

■ **48. (B) is correct.** The flow of sap from the leaves to the other part of the plant body is driven by hydrostatic pressure that develops inside the sieve tube, as phloem unloading creates a high solute concentration at the source end of the sieve tube—in contrast to with a low solute concentration at the sink end. Water flows through the tube, because the pressure is greatest at the tube's source end.

■ **49. (C) is correct.** It is theorized that glycolysis was occurring in the earliest cells on Earth. Oxygen did not start accumulating on Earth until long after prokaryotes appeared, so it is thought that they used glycolysis to generate ATP, since this process does not require oxygen. Also, given that glycolysis takes place in the cytosol, early prokaryotes (which lack membrane-bound organelles) could have carried it out.

■ **50. (D) is correct.** Regulatory genes are those that code for a protein—either a repressor or an inducer—that controls the transcription of another gene or a group of genes. In this case, allolactose is an inducer. When it is present in the cell, it causes the genes that transcribe for B-galactosidase to be turned on and to produce the enzyme. This represents an example of an inducible operon.

■ **51. (D) is correct.** Scientists can use the rate of genetic recombination between two genes in order to create a genetic map, which is an ordered list of the genetic loci along the length of a chromosome. This is due to the fact that the closer two linked genes are (linked genes are those located on the same chromosome), the less likely it is that they will undergo crossing over. The rate of recombination is proportionate to the distance between genes on a chromosome.

52. (D) is correct. The opening of the stomata in the leaves of a plant results in CO_2 being taken in from the atmosphere by the leaf. This, in turn, results in photosynthesis on a sunny day when light energy can also be captured by the leaf. Another result is in an increase in transpiration, or water loss by the leaf, through evaporation through the stomata.

53. (E) is correct. All of the answer choices are proof of the process of evolution except the existence of homologies among different species' diets. Since all living organisms require certain nutrients in order to survive—and since there are only so many consumable organic substances on Earth—the fact that two species might have similar components in their diet does not necessarily imply evolutionary relatedness.

54. (D) is correct. Since the frequency of the recessive allele is 0.35, we know that the frequency of the other allele is 1 - 0.35, which is equal to 0.65. The Hardy-Weinberg equation states that—if a population contains just two alleles for a given trait, and if the frequency of one of the alleles is known—the frequency of the other allele can be calculated using the equation $p + q = 1$. If you designate the frequency of the occurrence of the recessive allele as q, and use its value of 0.1, you can rearrange the equation to read $p + 0.35 = 1$. Then, $1 - 0.35 = 0.65$, which is equal to p, or the frequency of the other (in this case, the dominant) allele.

55. (A) is correct. The biologist has collected a large amount of protists from the pond; these organisms might be Euglena, for example. Most protists are unicellular, and they are the simplest eukaryotes. In this case, the protist is photoautotrophic, which means that it must have chloroplasts (because it carries out photosynthesis). Since these protists (algae) are plant-like, the cells are likely to have a central vacuole, too.

56. (C) is correct. This is an autosomal recessive trait. If it were sex-linked, it would be expressed in only one of the sexes—usually the male, since males have only one X chromosome. We know it is recessive because it does not appear in every generation; only dominant traits appear in every generation. Also, the original parents do not show the trait, but they have children who do.

57. (B) is correct. In facilitated diffusion, the transport of certain substances across a membrane is aided by transport proteins that span the membrane. These transport proteins are specialized for the solute that they transport. In the case of chemiosmosis, an H^+ gradient is established across the inner mitochondrial membrane by an ATP synthase, which pulls H^+ ions against their concentration gradient into the mitochondrial matrix and simultaneously synthesizes ATP.

58. (C) is correct. When protein enters the small intestine, the three enzymes that are responsible for breaking down proteins go to work. Trypsin and chymotrypsin break the peptide bonds between adjacent amino acids and are secreted by the intestinal epithelium in inactive form. They must be activated by another intestinal enzyme called enteropeptidase before beginning to break down the protein. Carboxypeptidase and aminopeptidase are two other enzymes that help to break down protein in the small intestine.

59. (C) is correct. This is an example of parasitic symbiosis, between the human and the tapeworm. The tapeworm gains nutrients from the human, whereas the human is harmed by the symbiotic relationship. The tapeworm causes intestinal blockage when it grows to its maximal length, and it can rob nutrients from the human to the extent that the human can develop nutritional deficiencies.

60. (D) is correct. The benthic zone in the ocean is the bottom of all aquatic zones; it is made up of sand and organic nutrients and is occupied by bacteria, fungi, seaweeds, algae, invertebrates, and some fishes. These organisms receive nutrients from the detritus that rains down from the ocean levels above it, where animals are dying and producing various metabolic wastes.

61. (B) is correct. Active transport involves the movement of substances across membranes against their concentration gradient. In this process, the cell must expend energy in the form of ATP. One example of active transport in the cell is the sodium-potassium pump, in which three sodium ions are pumped out of the cell and two potassium ions are pumped in. In the process ATP is dephosphorylated—it transfers one of its phosphate groups to the transport protein.

62. (A) is correct. Chimpanzees are thought to be the closest evolutionary relative of humans speaking. Chimpanzees and humans evolved from a common ancestor, and both are primates.

63. (B) is correct. An active skeletal muscle cell would be the most likely to have a high concentration of mitochondria in its cytoplasm, because mitochondria are the sites of cell respiration, and cell respiration is the source of ATP in the cell. Muscle cells use ATP in the process of contraction, and since they store only enough ATP for a few contractions, they must have many functional mitochondria to keep up the flow of ATP production when muscle contraction is continuous.

64. (C) is correct. The two most closely related organisms are the lobster and the spider. Both are arthropods, which are characterized by having segmented bodies, exoskeletons, and jointed appendages. Both organisms also have an open circulatory system.

65. (D) is correct. The activation energy of a reaction is the initial energy investment required in order for the reaction to proceed. It is the energy required in order to break the bonds of the substrate enough for the substrate to reach the highly unstable transition state. You can tell that answer *D* is the reaction energy of the uncatalyzed reaction (because the presence of a catalyst would decrease the overall energy of the reaction), so the taller curve must be the uncatalyzed reaction.

66. (C) is correct. The activation energy of the catalyzed reaction is represented by answer *C*. The overall energy of this reaction is significantly lower than that of the uncatalyzed reaction. This is because enzymes speed up the course of reactions by lowering the energy of activation so that the transition state is much easier to reach.

■ **67. (A) is correct.** The y-axis of this graph represents the free energy of the reaction. In energy profiles such as these, the y-axis always represents the free energy and the x-axis represents the progress of the reaction. This is an exergonic reaction, because the free energy of the reactants is higher than that of the products. This means that energy is given off in the course of the reaction.

■ **68. (B) is correct.** The transition state of a reaction is the highest-energy, most unstable form of the reactants in the reaction. The energy put into the reaction in order to make it "go"—also known as the activation energy—must be sufficient to enable the reactants to reach this transition state.

■ **69. (A) is correct.** The ovules are structures that develop in the plant ovary, and they contain the female gametophyte.

■ **70. (C) is correct.** The anther is the site of pollen production in the plant, and pollen grains contain the immature male gametophyte of a plant.

■ **71. (B) is correct.** The stigma is the sticky structure located at the end of the carpal. It is responsible for catching pollen grains.

■ **72. (E) is correct.** The sepals are usually green, and they are a whorl of modified leaves that enclose and protect the flower bud before it opens.

■ **73. (D) is correct.** The style is the stalk of the carpal of a flower; the ovary is at the base of the stalk; and the stigma is at the top of the style.

■ **74. (B) is correct.** Growth hormone (GH) is secreted by the anterior pituitary gland and affects many different target tissues. It promotes growth and stimulates the production of growth factors.

■ **75. (A) is correct.** Follicle-stimulating hormone, or FSH, is secreted by the anterior pituitary. It stimulates the production of ova and sperm in the gonads.

■ **76. (D) is correct.** Androgens are the male sex hormones, and the main androgen is testosterone. Androgens are synthesized in the testes, and they stimulate the development and maintenance of the male reproductive system.

■ **77. (C) is correct.** Melatonin is secreted by the pineal gland (in the brain); it is a modified amino acid that is secreted at night. The amount of melatonin secreted depends on the length of the night.

■ **78. (C) is correct.** The tundra is characterized by having a permafrost, which is a permanently frozen subsoil, and bitterly cold temperatures. It is extremely windy, so plants don't grow to be very tall in the tundra, and winters are long and dark.

■ **79. (D) is correct.** The chaparral is home to many dense, spiny evergreen bushes. The summers are long, hot, and dry, and the winters are mild and rainy. Plants in the chaparral are adapted for the periodic fires that ravage these biomes.

■ **80. (B) is correct.** The taiga is the coniferous forest biome, and these areas are characterized by frequent snowfall, harsh winters, short summers, and the presence of evergreens.

■ **81. (A) is correct.** Savannas are home to large grazing herbivores and their predators. They are covered in tall grasses with sporadic groups of trees. There is a considerable rainy season interrupted by periods of seasonal drought in areas containing savannas.

■ **82. (D) is correct.** Ribosomes are cell organelles that are constructed in the nucleolus and function as the site of protein synthesis in the cytoplasm.

83. (E) is correct. DNA replication occurs in the nucleus of the cell. The genetic material is replicated prior to mitotic or meiotic cell division.

84. (A) is correct. The electron transport chain consists of a number of molecules, mostly proteins, built into the inner membrane of a mitochondrion. Electrons removed from food are carried by NADH to the "top" end of the transport chain.

85. (C) is correct. Light reactions generate ATP by powering the addition of a phosphate group to ADP, a process called photophosphorylation. In the chloropast, the thylakoid membranes are the sites of the light reactions.

86. (B) is correct. Glycolysis takes place in the cytoplasm of the cell. In glycolysis, glucose is split into two molecules of pyruvate. This metabolic pathway occurs in all living cells, and it is the starting point for fermentation or cell respiration.

87. (E) is correct. The phylum Chordate contains two groups of invertebrates plus all of the animals with backbones. All chordates possess a notochord, a dorsal hollow nerve chord, pharyngeal slits, and a postanal tail as an embryo.

88. (C) is correct. Nematodes are found in aquatic habitats and have unsegmented bodies, with a tough exoskeleton called a cuticle. They have a complete digestive tract but lack a circulatory system, and they reproduce sexually.

89. (A) is correct. Rotifers have a complete digestive tract, with a separate mouth and anus, and a ring of cilia around their mouths which draws in water. They are pseudocoelomates.

90. (B) is correct. Porifera are sponges that have a sac-like body and are suspension feeders with no nerves or muscles. They draw water into a central cavity and filter it for nutrients. Sponges are also hermaphrodites.

91. (D) is correct. Platyhelminthes are flatworms that live in marine environments and other wet habitats. They include many parasitic species, and they have a gastrovascular cavity with just one opening. They are also acoelomates.

92. (A) is correct. Telomeres are regions found at the tips of chromosomes, and they are not made up of genes, but of repeating short sequences of DNA. These parts of chromosomes are copied by a special enzyme called telomerase.

93. (E) is correct. DNA ligase is an enzyme that is necessary for the replication of DNA; it catalyzes the covalent bonding of the 3' end of the new DNA fragment to the 5' end of the growing chain.

94. (B) is correct. DNA polymerase is another enzyme involved in DNA replication—it catalyzes the elongation of new DNA at the replication fork by adding nucleotides to the existing chain.

95. (C) is correct. Helicase is an enzyme that untwists the double helix of DNA at replication forks prior to DNA replication.

96. (D) is correct. The plant cell is probably in G_0 phase (G_0 phase is a nondividing phase). In many cells, there exists a G_1 checkpoint, and if at this checkpoint the cell is made to exit the cycle, it enters this nondividing G_0 phase. Some cells are perpetually in this phase, and since this plant cell has spent no time in any other phase besides the G_1, it is most likely arrested in G_0.

97. (C) is correct. The process of mitosis in the monkey liver cell took 18 minutes. The M phase of the cell cycle is the mitotic phase, and it is the phase in which the cell partitions the cytoplasm and organelles, plus the newly replicated DNA to two new daughter cells.

98. (B) is correct. According to this data, it took the *E. coli* cell 36 minutes to copy all of its DNA. The S phase is the phase in which all of the DNA in a cell's nucleus is copied, and during this time the cell is also growing—as it also is during the G_1 and G_0 phases.

99. (A) is correct. The data from this graph tells you that the genome of *E. coli* is much larger than the genome of the monkey; the cell spends much more time in the S phase, replicating its DNA, which means that there is probably much more of it to replicate.

100. (D) is correct. The mesoderm eventually gives rise to most organs and tissues in the body, including the kidney, heart, and inner layer of the skin (including mucous membranes). Therefore, mesoderm would have given rise to the heart, which was stained orange, and the mucous membranes, stained green in this example.

101. (B) is correct. The tissue stained blue was the brain, and the brain is derived from the ectoderm. Also arising from ectoderm is the rest of the nervous system, and the outer epidermal layer of our skin.

102. (D) is correct. The genes most likely to travel together and end up in the same daughter cell are W and E. This is because they are located close together on the chromosome. The closer two genes are on the chromosome, the less likely it is that crossing over will occur between them—and that they would be recombined (which would result in them being split up).

103. (D) is correct. If the rate of recombination between A and W is 5%, and the distance between gene A and gene W is 5 map units, and the distance between gene W and gene G is 15 map units, then you can calculate the rate of recombination between W and G by multiplying the 5% by 3, to get a 15% recombination rate.

104. (A) is correct. Species A is autotrophic and the primary producer of the ecosystem. This species is capable of capturing solar energy from the sun and converting it into the chemical energy contained in the bonds of organic compounds, which are used by the other animals in this ecosystem. It also is the only one that has arrows flowing only *from* it, indicating that it is not a consumer.

105. (B) is correct. Species B and C represent primary consumers—they consume only the autotrophs in this ecosystem, which are the primary producers. They are presumably herbivores, since their only food source is the primary producer.

106. (C) is correct. Species E consumes species A (presumably a plant), species C (presumably an animal), and species B (also presumably an animal). This means that species E is an omnivore—it eats both plants and animals. Species E is also both a primary consumer (since it consumes species A) and a secondary consumer (since it consumes species B and C).

107. (A) is correct. Both of the organisms that have lungs—both of the terrestrial animals, extract a higher percentage of oxygen from the air than do either of the animals that rely on tracheal systems for respiration. This enables you to conclude that lungs are more efficient.

108. **(A) is correct.** The most likely of these organisms to have hemolymph as its main circulatory fluid are the smallest organisms, which are presumably insects. These insects have open circulatory systems, in which no distinction is made between blood and interstitial fluid, and hearts that pump the hemolymph directly into the sinuses, which are open cavities for chemical exchange.

109. **(E) is correct.** In all of these animals, hemoglobin (an iron-containing molecule) is used to transport oxygen. It is contained in the red blood cells, or erythrocytes, which are a component of the blood of each of these animals.

110. **(B) is correct.** The two organisms in this chart in which gas exchange can occur without movement of some part of the animal are the small insect (#1) and the fish (#3). If insects are small enough, gas exchange simply takes place across the moist membranes of their trachea. Some fish can sit still in water and have the water flow across their gills, with gas exchange taking place.

111. **(D) is correct.** The DNA fragments migrated along the gel at rates according to their size—the smaller DNA fragments migrated more quickly through the dense gel and can be found near the bottom of the gel, whereas the larger fragments migrated more slowly and can be found near the top.

112. **(D) is correct.** One thing we can say about the DNA samples in this gel is that sample 2 must have been cut at more restriction sites than was sample 4, because more DNA fragments of different sizes were produced. This is evidenced by the greater number of bands on the gel in the lane of sample 2.

113. **(A) is correct.** The purpose of radioactively labeling these DNA samples was to make them visible when the gel was done running. After the gel is finished, and the DNA samples have migrated to a sufficient position to be distinguishable, the radiation is detected by radiography.

114. **(B) is correct.** The solution on Side A is hypotonic to the solution on Side B—it is less concentrated than the solution on Side B, at the time this experiment began.

115. **(B) is correct.** After two hours, the amount of glucose on side B will have increased. The membrane separating the two sides allows the passage of NaCl and not glucose, so there will be no movement of glucose—thus, no change in its concentration—but NaCl will travel down its concentration gradient to Side B.

116. **(C) is correct.** After two hours, the water column in Side B would be slightly higher. If the concentration of NaCl had equalized on both sides of the tube, the solution on side B would still be hypertonic to that on side A, thus water would flow through the membrane in an attempt to equalize its concentration on both sides of the tube, until gravitational pressures exerted an equal force to prevent it from rising further.

117. **(D) is correct.** The type of population growth seen in population A is exponential. Exponential population growth occurs under ideal conditions, and it means that the wild boar are reproducing at their reproductive capacity.

118. **(D) is correct.** The carrying capacity of population is about 1,500 members—it is at this point that the resources of the wild boar's environment start to become strained. The carrying capacity of an environment is defined as the

maximum population size that it can support at a particular time without being degraded.

■ **119. (A) is correct.** Density-dependent factors in a population cause it to level off at its carrying capacity. For instance, a death rate that rises as a population density rises is a density-dependent factor.

■ **120. (B) is correct.** From the graph, it looks as though the two species diverged as their population grew, through the process of speciation—whereby a new species was formed. This could have happened in many different ways. For example, two halves of the initial population could have been separated by natural disaster for a time long enough so that their gene pools became significantly different.

Free-Response Questions

1. (a) Negative feedback systems are very important in the body, especially in maintaining homeostasis, through factors like hormone secretion throughout the body. Negative feedback systems work much in the same way as thermostats do in houses. A receptor somewhere in the body detects a change in some factor in the animal's internal environment, and it transmits this information to a control center. The control center processes the information and directs a response to an effector, which carries out the response. In negative feedback, a change in the variable triggers the control center to prevent further change in the same direction.

(b) The birth control pill blocks the secretion of GnRH by the hypothalamus, and FSH and LH by the pituitary. Together, GnRH, LH, and FSH all work to create an elaborate feedback loop that syncs up the ovarian cycle and the menstrual cycle. During the ovarian cycle, the pituitary secretes FSH and LH in small amounts, and this stimulates the secretion of GnRH by the hypothalamus. FSH stimulates the follicle to grow, and the follicle cells secrete estrogen, which in negative feedback keeps the secretions of FSH and LH relatively low. Later in the cycle, the follicles begin to secrete estrogens rapidly, which has the effect of suddenly causing the increased secretion of FSH and LH. The sudden increase in the secretion of LH is what stimulates ovulation. By blocking the release of LH, the birth control pill prevents ovulation from occurring. Ovulation is the release of the egg from the ovaries, and this is the time in the ovarian cycle when a female can become pregnant if sperm is present.

This is a good free response answer because it shows working knowledge of the following terms:

negative feedback system	*GnRH*	*follicle*
hormone	*Hypothalamus*	*estrogens*
receptor	*FSH*	*ovarian cycle*
control center	*LH*	*menstrual cycle*
effector	*pituitary gland*	*ovulation*

This response also demonstrates an understanding of the following biological processes—a negative feedback system and the female menstrual/ovarian cycle.

2. (a) Three ways in which gene expression in a cell is controlled are through chromatin packing, DNA methylation, and histone acetylation. In chromatin packing, when the genetic material is in heterochromatin form, it is highly condensed and packed. This is because the proteins involved in transcription cannot get near enough to the DNA to do their jobs. In DNA methylation, methyl groups are attached to specific regions of DNA immediately after it is synthesized. In some cases, this is thought to be responsible for these genes' long-term inactivation. Finally, in histone acetylation, acetyl groups are attached to certain amino acids of histone proteins, and when the histones are acetylated, their shape alters so that they are less tightly bound to DNA; this enables the proteins involved in transcription to move in and begin work. When histones are deacetylated, DNA transcription is impossible.

(b) In order to tell how actively a certain cell is transcribing its DNA and translating its mRNA into protein, you could do a few things in the lab. Firstly, you could monitor the rate of relaxation of the heterochromatin in the cell nucleus—the more relaxed the chromosomes are, the more DNA is being transcribed. Secondly, you could monitor the amount of uptake of cytosine, guanine, uracil, and adenine in the cell—the rate at which these bases are taken up would be an indicator of the rate at which they are being incorporated into mRNA in transcription. Finally, you could monitor the rate at which free amino acids are being consumed in the cell. This would indicate the rate at which they are being incorporated into growing peptide chains in translation.

This is a good free-response answer because it shows working knowledge of the following key terms:

chromatin packing	*transcription*	*uracil*
DNA methylation	*heterochromatin*	*thymine*
histone acetylation	*cytosine*	*amino acids*
histones	*adenine*	*peptides*

The response also shows an understanding of the following important biological processes—control of gene expression in eukaryotes and the process of transcription and translation.

3. (a) Humans, gibbons, orangutans, and chimpanzees are primates, so all have an opposable thumb and feet that can grip. The primates are grouped into the prosimians and the anthropoids. All four of these species are anthropoids, and the gibbons, orangutans, humans, and chimpanzees are all hominoids. It is thought that humans and chimps are two divergent branches of the hominoid

tree that evolved from a common ancestor (that was neither human nor chimp) about 5 million years ago.

(b) One of the three kinds of evidence that was used to determine relationships among these species is fossil evidence. Fossils are impressions or parts of organisms from the past that are preserved in rock. Most fossils are found in sedimentary rock, and the age of fossils can be determined to a certain extent by the depth of the rock layer in which they are found. Another type of evidence that was used to determine relationships among the arthropods is the existence of homologies. Homologies are shared characteristics that are the result of two species having evolved from a common ancestor. The third type of evidence is homologies in DNA—molecular homologies. The likeness of two species' DNA tells how closely they are related evolutionarily.

(c) Some of the characteristics of the ancestors of *Homo sapiens* are a relatively large brain (which is associated with the use of language and with other cultural aspects), a longer jaw and certain resulting changes in the teeth, bipedal posture, and a reduced difference in the sizes of the two sexes—males and females were more nearly the same size.

This is a good free-response answer because it shows a working knowledge of the following key terms:

primates   *fossils*
prosimians   *homologies*
arthropods

The response also shows an understanding of these key biological concepts—evolution, the relatedness of primates, how humans are similar and different from other primates.

4. (a) In the morning, if it were a sunny day that was not too hot, the stomata in the leaves of the plant would open and allow the intake of carbon dioxide and the exit of oxygen, in order for the plant to begin photosynthesis. The plant would then begin to use the light energy from the sun to turn carbon dioxide into sugar molecules, which it would use as food, and oxygen. The process of transpiration is a critical one; when the stomata are open, too, transpiration is the loss of water through the stomata, and the plant must carefully balance taking in carbon dioxide with losing too much water. Therefore, the plant must keep its stomata open during the day in order to use the sun's energy to photosynthesize, but it runs the risk of losing too much water through transpiration. When night falls, the plant will close its stomata to prevent unnecessary water loss. Since it can no longer get energy from the sun for photosynthesis, so doesn't need carbon dioxide.

Plants are generally bound to follow a biological clock that controls their circadian rhythms. The amounts of transpiration and enzyme synthesis fluctuate during the course of the day. Some of this is in response to changes in humidity and temperature that occur during the course of the day, but even without those external changes, a plant knows how long the day is because of its biological clock. Since this is a flowering plant, it will measure the length of the night and will flower when the night length reaches a critical length; this how a plant determines the time of season.

(b) If this plant is moved to a window that faces west, plant hormones will act to start its growth in the direction facing the sun. This is thought to be because a plant responds to light by an asymmetrical distribution of auxin going down from the tip of the plant, which causes the cells on the darker side of the plant to elongate (not divide) more than the cells on the brighter side of the plant. Growth of a plant toward a light source is known as phototropism.

This is a good free-response answer because it shows knowledge of the following terms:

stomata	*circadian rhythm*
transpiration	*auxin*
photosynthesis	*phototropism*
biological clock	

The response also shows an understanding of the following important biological processes—the daily metabolic cycle of plant and plant responses to light.

Practice Test 2

Biology
Section 1

Time—1 hour and 30 minutes

Directions: Each of the questions or incomplete statements below is followed by five suggested answers or completions. Select the one that is best in each case and then fill in the corresponding oval on the answer sheet.

1. A geneticist crosses two rabbits, both of which have brown fur. In rabbits, brown fur is dominant over white fur. Six of the eight offspring produced have brown fur, and the other two have white coats. The genotypes of the parents were most likely which of the following?
   (A) $BB \times bb$
   (B) $BB \times Bb$
   (C) $Bb \times bb$
   (D) $Bb \times Bb$
   (E) $bb \times bb$

2. A biologist isolates an organism from a pond and describes it in his lab notebook as having both a larval and an adult stage, air-breathing lungs, the digestive system of a carnivore, eggs that lack shells, and camouflage epidermis that secretes toxins, and as exhibiting external fertilization. This organism is probably which of the following?
   (A) A snake
   (B) A frog
   (C) A freshwater clam
   (D) A salamander
   (E) A lungfish

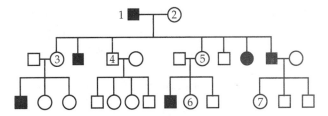

3. In the pedigree above, the squares represent males and the circles represent females. Individuals who express a particular trait are represented by shaded figures. Which of the following patterns of inheritance best explains the transmission of the trait?
   (A) Sex-linked dominant
   (B) Sex-linked recessive
   (C) Autosomal dominant
   (D) Autosomal recessive
   (E) Incompletely dominant

4. Which of the following is an example of simple diffusion across a membrane?
   (A) The movement of H⁺ across the thylakoid membrane during photosynthesis
   (B) The uptake of neurotransmitters by the postsynaptic membrane during the transmission of a nerve impulse
   (C) The movement of oxygen in the alveoli across the epithelial membrane and into the bloodstream
   (D) The exchange of sodium and potassium across a cell membrane through the Na^+/K^+ pump
   (E) The movement of glucose across the body cell membranes and the cells of the liver, which stores it as glycogen

5. Which of the following correctly represents the order of the tissues through which water and minerals will pass on their way up from a plant's roots?
   (A) Root hair, endodermis (Casparian strip), epidermis, cortex, stele
   (B) Root hair, cortex, epidermis, endodermis (Casparian strip), stele
   (C) Root hair, stele, cortex, endodermis (Casparian strip), epidermis
   (D) Root hair, epidermis, stele, cortex, endodermis (Casparian strip)
   (E) Root hair, epidermis, cortex, endodermis (Casparian strip), stele

6. Which of the following evolved before algae?
   (A) Bacteria
   (B) Hydra
   (C) Cnidarians
   (D) Fungi
   (E) Protists

7. Thinking that you will go for a swim, you approach a swimming pool and submerge your toes into the water. The water is very cold, which prompts you reflexively to withdraw your foot. In terms of a feedback circuit, in this case the muscles that moved in order to allow you to withdraw your foot quickly can be thought of as which of the following?
   (A) The receptor
   (B) The control center
   (C) The effector
   (D) The signaler
   (E) The sensor

8. Both insertions and deletions cause which of the following types of mutation?
   (A) Missense mutation
   (B) Nonsense mutation
   (C) Gene substitution
   (D) Base-pair substitution
   (E) Frameshift mutation

9. The stomata—the openings on the underside surface of a plant leaf through which carbon dioxide is taken up and oxygen is expelled— are opened up as a result of
   (A) movement of mesophylls away from the stomatal opening
   (B) increased turgidity in the guard cells
   (C) decreased turgidity in the guard cells
   (D) growth of the guard cells toward the mesophyll
   (E) elongation of the guard cells toward the mesophyll

10. All of the following are characteristic of a population in Hardy-Weinberg equilibrium EXCEPT
    (A) the population must be very large
    (B) there must be no migration into or out of the population
    (C) the members of the population must be randomly mating
    (D) there must be only two alleles present for each characteristic in the population
    (E) natural selection must not be operating in the population

11. The region in the cell from which the mitotic spindles radiate out is called which of the following?
    (A) Centromere
    (B) Centrosome
    (C) Cytoplasm
    (D) Kinetochore
    (E) Metaphase plate

12. In plants that undergo alternation of generations, the gametophyte stage is always
    (A) a large visible plant
    (B) a seed
    (C) diploid
    (D) haploid
    (E) unicellular

13. Bacteria reproduce by which of the following processes?
    (A) Mitosis
    (B) Meiosis
    (C) Binary fission
    (D) Binary division
    (E) Cleavage

14. All of the following are factors contributing to the ascent of water through the xylem in plants EXCEPT
    (A) transpiration
    (B) low water potential at one end
    (C) cohesion of water to the vessel walls
    (D) adhesion of water to the vessel walls
    (E) sources and sinks

15. In plants, the abscission, or dropping, of leaves is triggered by changes in
    (A) cytokinin
    (B) ethylene
    (C) abscisic acid
    (D) gibberellins
    (E) brassinosteroids

16. Near the lungs, a branch from the pulmonary artery would contain which of the following?
    (A) Oxygen-rich blood
    (B) Oxygen-poor blood
    (C) Dissolved nutrients from the stomach
    (D) Blood rich in carbon monoxide
    (E) Lymph

17. Gel electrophoresis can be used for which of the following laboratory procedures?
    (A) Determining the molecular weight of proteins and nucleic acids
    (B) Determining the charge of proteins and nucleic acids
    (C) Separating nucleic acids and proteins on the basis of their size
    (D) Separating nucleic acids and proteins on the basis of their charge
    (E) Breaking up proteins and nucleic acids into their monomers

18. In humans, if red hair (R) is dominant to brown hair (r), and freckles (F) are dominant to no freckles, (f), what fraction of the progeny of the cross $RrFf \times RRff$ will have red hair and no freckles?
    (A) $\frac{9}{16}$
    (B) $\frac{1}{2}$
    (C) $\frac{3}{8}$
    (D) $\frac{3}{16}$
    (E) $\frac{1}{16}$

19. Which of the following can be observed best by using a compound light microscope?
    (A) Atoms and molecules
    (B) Proteins
    (C) Ribosomes
    (D) Bacteria
    (E) Viruses

GO ON TO THE NEXT PAGE

20. The phenomenon by which plants will bend toward or away from a light source is known as
    (A) photoaffinity
    (B) photogropism
    (C) phototropism
    (D) photoperiodicy
    (E) photophilia

21. All of the following are functions of micro-tubules in the cell EXCEPT
    (A) components of cilia, used for locomotion
    (B) components of flagellum, used for loco-motion
    (C) involved in the movement of chromo-somes during cell division
    (D) comprising the cytoskeleton, function in cell support
    (E) part of the nuclear membrane

22. Which of the following organelles is the site of macromolecule hydrolysis in the cell?
    (A) Mitochondria
    (B) Centrosome
    (C) Lysosome
    (D) Golgi apparatus
    (E) Peroxisome

23. Which of the following describes how a dog that is prodded while asleep will respond to the touch initially, but will then ignore the touch after repeated prodding?
    (A) Habituation
    (B) Imprinting
    (C) Reasoning
    (D) Instinct
    (E) Trial and error

24. Which of the following is characteristic of a plant cell but not of an animal cell?
    (A) Rough endoplasmic reticulum
    (B) Cell membrane
    (C) Ribosomes
    (D) Large central vacuole
    (E) Golgi apparatus

25. The process according to which eventually there are barriers to successful inbreeding between two species with separate geographic ranges is called
    (A) sympatric speciation
    (B) allopatric speciation
    (C) adaptive radiation
    (D) polyploid speciation
    (E) exaptation

26. The fact that pairs of alleles will segregate randomly during gamete formation describes which of the following laws?
    (A) The law of segregation
    (B) The law of independent segregation
    (C) The law of equal inheritance
    (D) The law of independent assortment
    (E) The law of equal segregation

27. In a savanna, which of the following organisms is the major primary producer?
    (A) Algae
    (B) Grasses
    (C) Trees
    (D) Insects
    (E) Small mammals

28. In cows, eye color is controlled by a single gene with two alleles. When a homozygous cow with brown eyes is crossed with a homozygous cow with green eyes, cows with blue eyes are produced. If the blue-eyed cows are crossed with each other, what fraction of their offspring will have brown eyes?
    (A) 0
    (B) ¼
    (C) ½
    (D) ¾
    (E) 1

29. Which of the following is not an adaptation for gas exchange?
    (A) Lungs
    (B) Tracheal system
    (C) Gills
    (D) Moist epidermis
    (E) Sinuses

30. It is theorized that birds evolved from which of the following groups?
    (A) Fish
    (B) Amphibians
    (C) Reptiles
    (D) Mammals
    (E) Prokaryotes

31. Which of the following best characterizes the reaction represented below? $A + B \rightarrow AB +$ energy
    (A) Exergonic reaction
    (B) Endergonic reaction
    (C) Oxidation-reduction reaction
    (D) Catabolism
    (E) Hydrolysis

32. During prophase of mitosis, DNA in the nucleus exists in which of the following forms?
    (A) Daughter chromosomes
    (B) Chromatin
    (C) Chromosomes consisting of two sister chromatids
    (D) Single sister chromatids
    (E) Single linear chromosomes

33. One way to measure the metabolic rate of a cell would be to measure the rate at which
    (A) CO_2 is consumed by the cell
    (B) O_2 is consumed by the cell
    (C) water is consumed by the cell
    (D) O_2 is produced by the cell
    (E) CO_2 is produced by the cell

34. Which of the following is the site of translation in the cell?
    (A) The nucleus
    (B) The Golgi apparatus
    (C) Smooth ER
    (D) Rough ER
    (E) Mitochondria

35. In certain plant cells, the synthesis of ATP occurs in which of the following?
    (A) Ribosomes and mitochondria
    (B) Ribosomes and chloroplasts
    (C) Mitochondria and chloroplasts
    (D) Mitochondria and the cytoplasm
    (E) Chloroplasts and the cytoplasm

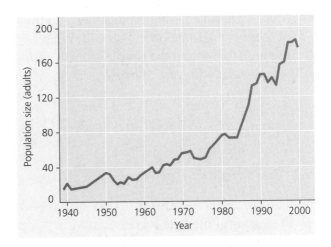

36. The graph above shows the rate of growth of a population of squirrels in a certain geographic area in Connecticut over the course of the last 60 years. This population is exhibiting which of the following types of growth?
    (A) Logistic growth
    (B) Probable growth
    (C) R-selected growth
    (D) K-selected growth
    (E) Exponential growth

GO ON TO THE NEXT PAGE

37. Genes M and N are located on different chromosomes, and the probability of their undergoing crossing over is quite low. If the probability of allele M segregating into a gamete is ⅙, and the probability of allele N segregating into a gamete is ¼, then the probability that both of them will segregate into the same gamete is
(A) ¹⁄₁₂
(B) ¼
(C) ⁵⁄₁₂
(D) ¾
(E) 1

38. Two individuals, both of whom are carriers for cystic fibrosis (a recessively inherited disorder), produce 3 children together, and none of these children has cystic fibrosis. What is the probability that the couple's fourth child will be born with cystic fibrosis?
(A) 0%
(B) 25%
(C) 50%
(D) 75%
(E) 100%

39. The overall temperature of the earth has been increasing along with increases in the amount of what gas?
(A) Oxygen
(B) Nitrogen
(C) Hydrogen
(D) Carbon monoxide
(E) Carbon dioxide

40. Which of the following groups comprise a strand of DNA?
(A) Phosphate groups, deoxyriboses, and nitrogenous bases
(B) Phosphate groups, riboses, and nitrogenous bases
(C) Phosphate groups, deoxyriboses, and amino acids
(D) Phosphate groups, riboses, and amino acids
(E) Deoxyriboses and nitrogenous bases

41. The statement that evolutionary changes are composed of rapid bursts of speciation that alternate with long periods in which species do not change significantly is known as
(A) gradualism
(B) punctuated gradualism
(C) punctuated equilibrium
(D) sympatric speciation
(E) allopatric speciation

42. Which of the following vertebrates lacks an amnion during its development?
(A) Bird
(B) Human
(C) Lizard
(D) Frog
(E) Alligator

43. Which of the following is capable of reverse transcription, with an RNA → DNA information flow?
(A) Viruses
(B) Retroviruses
(C) T cells
(D) B cells
(E) H cells

44. Compared with prokaryotic cells, eukaryotic cells are generally
(A) smaller but more complex
(B) larger and more complex
(C) smaller and less complex
(D) larger but less complex
(E) the same size but more complex

45. Which of the following plant hormones is responsible for stimulating stem elongation, root growth, and cell differentiation?
(A) Ethylene
(B) Abscisic acid
(C) Cytokinin
(D) Gibberellin
(E) Auxin

46. The direct manipulation of genes for practical purposes is known as
    (A) recombination
    (B) gene cloning
    (C) genetic engineering
    (D) DNA engineering
    (E) genomics

47. In terms of evolution, which of the following is closer to fungi?
    (A) Plants
    (B) Animals
    (C) Archae
    (D) Bacteria
    (E) Viruses

48. In humans, which of the following glands is responsible for secreting several hormones involved in the human reproductive cycle?
    (A) Thyroid gland
    (B) Adrenal cortex
    (C) Adrenal medulla
    (D) Anterior pituitary
    (E) Posterior pituitary

49. Evolutionarily, which of the following was thought to be the first genetic material?
    (A) DNA
    (B) RNA
    (C) cDNA
    (D) mRNA
    (E) tRNA

50. When a break in the epidermal layer of humans occurs, which of the following types of blood cells travels in great number to the break to release clotting factors?
    (A) Leukocytes
    (B) Erythrocytes
    (C) Helper T cells
    (D) Helper B cells
    (E) Platelets

51. In photosynthesis, the functional product of the light reactions of photosynthesis
    (A) are ATP and NADPH
    (B) are ATP and NADH
    (C) is glyceraldehyde
    (D) is glucose
    (E) are carbohydrates

52. Which of the following groups is best characterized as being eukaryotic and saprophytic and as having hyphae?
    (A) Monera
    (B) Plantae
    (C) Virus
    (D) Fungi
    (E) Animalia

53. In humans, hemophilia is a sex-linked recessive trait. If a man and a woman produce a son who is a hemophiliac, which of the following must be true?
    (A) The father is a hemophiliac.
    (B) Both parents carry the allele for hemophilia.
    (C) Neither parent carries the allele for hemophilia.
    (D) The father carries the allele for hemophilia.
    (E) The mother carries the allele for hemophilia.

54. If a horse breeds with a donkey, a mule is produced. Mules are not capable of breeding with either parental species, or each other, to produce offspring. This is an example of what type of postzygotic barrier?
    (A) Reduced hybrid viability
    (B) Hybrid sterility
    (C) Hybrid breakdown
    (D) Mechanical isolation
    (E) Gametic isolation

GO ON TO THE NEXT PAGE

55. Which of the following processes is carried out less efficiently by a C$_3$ plant than by a C$_4$ plant?
(A) Photon absorption
(B) Chemiosmosis
(C) Photolysis
(D) Carbon fixation
(E) The transport of sugars

56. Radioactive isotopes can be used to date fossils. The amount of time it takes for half of a radioactive isotope to decay is also known as the substance's
(A) release rate
(B) radioactive decay rate
(C) half-life
(D) time scale
(E) decay rate

57. The female gametophytes of a plant develop in the ovaries of the plant, whereas the male gametophyte develops in which of the following plant structures?
(A) Stigma
(B) Style
(C) Carpel
(D) Anther
(E) Sepal

58. Which of the following is the most direct result of the presence of salivary amylase in the mouth?
(A) The breakdown of proteins
(B) The breakdown of polypeptides
(C) The breakdown of lipids
(D) The breakdown of carbohydrates
(E) The breakdown of nucleic acids

59. Which of the following classes of animals exists in the greatest number on earth?
(A) Monera
(B) Insecta
(C) Archae
(D) Bacteria
(E) Protista

60. The leaves of a plant appear green to us because
(A) chlorophyll reflects green light
(B) chlorophyll absorbs green light
(C) chlorophyll reflects red light
(D) chlorophyll reflects blue light
(E) chlorophyll is green, and plants contain hundreds of chlorophyll molecules

61. The fact that ice is less dense than liquid water and the fact that water is a good insulator are due to what type of bonds that form between water molecules?
(A) Ionic
(B) Covalent
(C) Polar covalent
(D) Hydrogen
(E) Double

62. Which of the following is the insulating layer wrapped around nerve cells that acts to speed up the transmission of nerve impulses?
(A) Axons
(B) Dendrites
(C) Synaptic terminal
(D) Myelin sheath
(E) Nodes of Ranvier

63. Which of the following is the substrate in the Krebs cycle?
(A) Carbon dioxide
(B) Acetyl-CoA
(C) Citrate
(D) Oxaloacetate
(E) Glucose

64. Insects, spiders, and crustaceans are all a part of which phylum?
(A) Arthropoda
(B) Annelida
(C) Chordata
(D) Nemertea
(E) Cnidaria

Directions: Each group of questions below consists of five lettered headings (or five lettered items in a graph) followed by a list of numbered phrases or sentences. For each numbered phrase or sentence select the one heading (or item) that is most closely related to it and fill in the corresponding oval on the answer sheet. Each heading may be used once, more than once, or not at all in each group.

Questions 65–68
    (A) Meisosis II
    (B) Meiosis I
    (C) Binary fission
    (D) Mitosis
    (E) Interphase

65. The process during which a cell's nucleus is divided

66. The process during which prokaryotes reproduce

67. The process during which the diploid chromosome number is halved

68. The process during which the genetic material of the cell is replicated

Questions 69–73
    (A) Amphibia
    (B) Reptilia
    (C) Echinodermata
    (D) Chordata
    (E) Chondrichthyes

69. Members have cartilaginous skeletons and include sharks and sea rays.

70. Members have a water vascular system and include sea stars and sea cucumbers.

71. Members have eggs without shells and include frogs and salamanders.

72. Members have scales and lungs and include snakes and lizards.

73. Members have a notochord and pharyngeal slits and include humans.

Questions 74–78
    (A) Electron transport chain
    (B) Chemiosmosis
    (C) Glycolysis
    (D) The Krebs Cycle (citric acid cycle)
    (E) Light reactions of photosynthesis

74. Drives the synthesis of ATP through a hydrogen ion gradient

75. Occurs in all living cells and is the starting point for aerobic respiration or fermentation

76. Is part of cell respiration and completes the breakdown of glucose into carbon dioxide

77. Shuttle electrons and releases energy that is used to make ATP

78. Solar energy is converted to chemical energy.

Questions 79–82
    (A) Population
    (B) Community
    (C) Species
    (D) Niche
    (E) Biome

79. Members are capable of interbreeding and are anatomically similar.

80. The biotic and abiotic resources a species uses in its environment

81. Individuals of one species that live in a discreet geographic area

82. All the organisms that live within a discreet geographic area

GO ON TO THE NEXT PAGE

Questions 83–87

(A) Antigens
(B) Antibodies
(C) Histamines
(D) Eosinophils
(E) Macrophages

83. Large phagocytotic cells that engulf microbes

84. A type of white blood cell

85. Proteins that bind antigens

86. Foreign molecules that elicit an immune response

87. Chemical signals released in response to injury

Questions 88–92

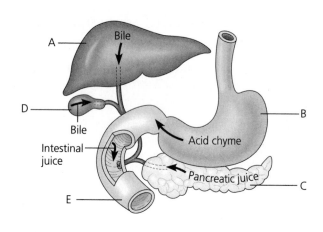

88. Stomach

89. Gall bladder

90. Duodenum

91. Pancreas

92. Liver

Questions 93–95

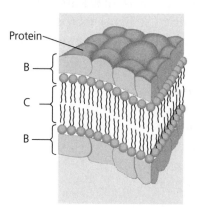

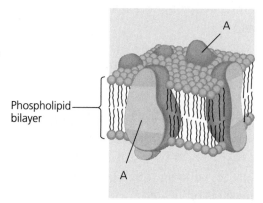

93. The hydrophilic zone of the plasma membrane

94. The hydrophobic zone of the plasma membrane

95. A membrane protein

Directions: Each group of questions below concerns an experimental or laboratory situation or data. In each case, first study the description of the situation or data. Then choose the one best answer to each question following it and fill in the corresponding oval on the answer sheet.

Questions 96–99
The rate of reaction for 3 enzymes was calculated at different temperatures, and the rate of reaction for 2 additional enzymes was calculated at different pHs; this data was graphed below. Assume that the y-axes share the same scale.

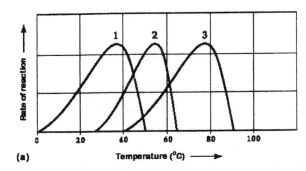

(a)

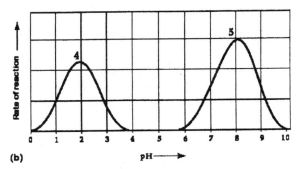

(b)

96. Which of the enzymes would be most likely to be able to function in the bloodstream of humans?
(A) 1 and 4
(B) 1, 2, and 4
(C) 1, 2, and 5
(D) 3 and 4
(E) 3 and 5

97. Which of the enzymes would work best in the small intestine of the human?
(A) 1
(B) 2
(C) 3
(D) 4
(E) 5

98. Which of the enzymes would be most likely to function in the hot springs of Yellowstone Park?
(A) 1
(B) 2
(C) 3
(D) 4
(E) 5

99. Which of these enzymes is most efficient— that is, has the highest rate of reaction?
(A) 1
(B) 2
(C) 3
(D) 4
(E) 5

Questions 100–103
A scientist wanting to study the mammalian heart is experimenting on a white rat. She injects different radioactive elements into different sections of the rat's heart and charts where she has injected each substance, as shown below.

Heart chamber	Radioactive Isotope used
Right ventricle	^{32}P
Right atria	^{3}H
Left ventricle	^{14}C
Left atria	^{238}U

100. Just after its injection, where would the radioactive isotope ^{238}U be detected first?
(A) Left ventricle
(B) Right ventricle
(C) Left atria
(D) Systemic capillaries
(E) Pulmonary capillaries

GO ON TO THE NEXT PAGE

101. Just after its injection, the blood injected with ^{14}C would be detected performing which of the following tasks in the body?
(A) Picking up oxygen from the systemic capillaries of the body
(B) Dropping off oxygen in the systemic capillaries of the body
(C) Picking up oxygen in the capillaries of the lungs
(D) Dropping off carbon dioxide in the capillaries of the lungs
(E) Dropping off oxygen in the capillaries of the lungs

102. Just after injection, where would the ^{3}H be detected?
(A) In the right ventricle
(B) In the right atria
(C) In the left ventricle
(D) In the systemic circulation
(E) In the pulmonary circulation

103. Just after its injection of ^{32}P, the blood injected with ^{32}P would be detected performing which of the following tasks in the body?
(A) Picking up oxygen from the systemic capillaries of the body
(B) Dropping off oxygen in the systemic capillaries of the body
(C) Picking up oxygen in the capillaries of the lungs
(D) Dropping off carbon monoxide in the capillaries of the lungs
(E) Dropping off oxygen in the capillaries of the lungs

Questions 104–106 refer to the graph shown below.

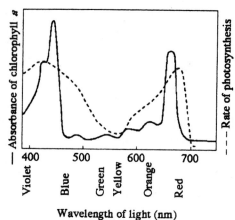

104. The graph shows that the rate of photosynthesis has a peak just before 700 nm of light. Which of the following is the most likely explanation for this fact?
(A) Light excites chlorophyll b optimally at this wavelength.
(B) Light excites carotenoids optimally at this wavelength.
(C) Photosystem I pigments have their optimal light absorption in this range.
(D) Photosystem II pigments have their optimal light absorption in this range.
(E) Both photosystems I and II pigments have their optimal light absorption in this range.

105. A biologist would use which of the following terms to refer to the solid line in the graph?
(A) Action spectrum
(B) Absorption spectrum
(C) Photostimulation curve
(D) Electromagnetic spectrum
(E) Visible light spectrum

106. Which of the following is the best reason why the curve for the absorbency of light by chlorophyll a does not perfectly match the rate of photosynthesis?
   (A) The rate of photosynthesis is always fractionally slower than the rate of absorbency by chlorophyll a.
   (B) The rate of photosynthesis is always fractionally faster than the rate of absorbency by chlorophyll a.
   (C) There are fewer chlorophyll a molecules in the cell than the other molecules involved in photosynthesis, so chlorophyll a is the rate-limiting reagent.
   (D) Chlorophyll a is not the only photosynthetically important pigment in chloroplasts.
   (E) Light of about 550 inhibits all photosynthesis.

Questions 107–110
An ecologist studying a certain biogeographic area has sketched the following food web for the community that lives there. The arrows represent energy flow, the letters represent species.

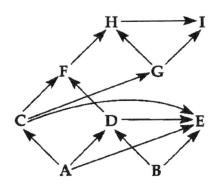

107. Which of the following is most likely to be autotrophic?
   (A) E
   (B) A
   (C) I
   (D) G
   (E) D

108. Which members of the food web are secondary consumers?
   (A) D, E, and F
   (B) A, B, and C
   (C) G, H, and I
   (D) E, F, and G
   (E) B, D, and E

109. Which of the species in the food web above are carnivores?
   (A) E, F, G, H, and I
   (B) F, G, H, and I
   (C) C, D, and E
   (D) F and G only
   (E) H and I only

110. If this is a savanna ecosystem, what organisms are species A and B most likely to represent?
   (A) Two species of small low-growing bushes
   (B) Two species of lichen
   (C) Two species of insects
   (D) Two species of rodent
   (E) Two species of grasses

Questions 111–113 refer to the following table, which shows the temperature at which the DNA of various species has been found to denature.

Species	Temperature at Which DNA Denatures
A	25°C
B	80°C
C	72°C
D	58°C
E	57°C

111. Which of the following probably has the most guanine-cytosine base pairs (as opposed to adenine-thymine base pairs) in its DNA?
   (A) A
   (B) B
   (C) C
   (D) D
   (E) E

GO ON TO THE NEXT PAGE

112. Which of the species in this table are most
likely to be related most closely evolutionarily?
(A) A and B
(B) B and C
(C) C and D
(D) D and E
(E) A and E

113. What other experimental method could be
used to denature DNA strands in order to
obtain the same type of information?
(A) Adding a buffer to the DNA samples
(B) Adding a solvent to the DNA samples
(C) Slowly lowering the pH of the DNA
samples
(D) Slowly adding free bases to the DNA
samples
(E) Slowly increasing the pH of the DNA
samples

*Questions 114–117 refer to the reaction A + B → C
+ D and to the graph of the free energy changes in
this reaction, as shown below.*

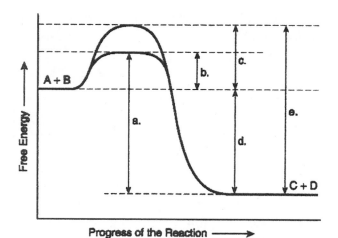

114. Which of the following represents the energy
of the transition state of the uncatalyzed
reaction?
(A) A
(B) B
(C) C
(D) D
(E) E

115. Which of the following terms best describes
this reaction?
(A) Exergonic
(B) Endergonic
(C) Oxidation-reduction
(D) Hydrolysis
(E) Complex

116. Which of the following represents the activa-
tion energy of the catalyzed reaction?
(A) A
(B) B
(C) C
(D) D
(E) E

117. Which of the following values would remain
the same regardless of whether this reaction
were catalyzed or not catalyzed?
(A) A
(B) B
(C) C
(D) D
(E) E

Questions 118–120 refer to the graph below, which shows the productivity of artichoke crops on a small organic farm in New Mexico. The temperature in this geographic area has been steadily increasing over the past thirty years.

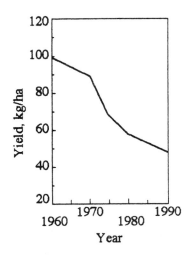

118. Which of the following would be the best advice for the farmer of this land?
(A) Replace your current crop with a potato crop.
(B) Replace your current crop with a corn crop.
(C) Replace your current crop with an asparagus crop.
(D) Use more fertilizer.
(E) Use less fertilizer.

119. One plant macronutrient that is often a component of fertilizer—since it cannot be taken up from the atmosphere and so must exist in the soil—is
(A) carbon
(B) oxygen
(C) hydrogen
(D) nitrogen
(E) iron

120. Most of the total mass of a plant comes from which of the following sources?
(A) N_2
(B) O_2
(C) CO_2
(D) H_2O
(E) H_2

END OF SECTION I

Biology
Section II

Time—1 hour and 30 minutes

Answer all questions. Number your answer as the question is numbered below.

    Answers must be in essay form. Outline form is NOT acceptable. Labeled diagrams may be used to supplement discussion, but in no case will a diagram alone suffice. It is important that you read each question completely before you begin to write.

1. The cell membrane is one of the most important parts of a cell; it allows the selective passage of materials into and out of the cell, thereby maintaining a constant, desired internal composition.
   (a) Discuss the components of a typical animal cell membrane, as well as the roles each of these components plays in regulating the cell's internal environment.
   (b) Discuss the ways in which the following can gain entrance to the cell, including
      • viral DNA;
      • hormones;
      • and water molecules.

2. It is thought that the terrestrial plants we see around us today have evolved from aquatic algae.
   (a) Discuss three obstacles to the evolutionary movement of plants onto land.
   (b) Discuss three adaptations that terrestrial plants have undergone in order to enable them to overcome the obstacles above, as well as any others you would like to include.

3. Describe the process of cell division in a typical plant cell, including:
   (a) the cell cycle of a plant cell;
   (b) the process of mitosis;
   (c) and cytokinesis.

4. It is theorized that, evolutionarily, glycolysis was the first metabolic pathway for producing ATP.
   (a) Give three pieces of evidence that might be used to argue the point above.
   (b) Describe how the Krebs cycle is related to chemiosmosis and to oxidative phosphorylation.

END OF EXAMINATION

ANSWERS AND EXPLANATIONS

Multiple-Choice Questions

▌**1. (D) is correct.** The ratio of brown-furred rabbits to white-furred rabbits was 3:1, and this means that we can deduce that the genotypes of the offspring were *BB*, *Bb*, *Bb*, and *bb*. By doing a Punnett square, you can find that the correct combination of alleles to produce offspring in that ratio is *Bb* × *Bb*.

▌**2. (B) is correct.** The organism that this biologist isolated is a frog, which is part of the class Amphibia and the order Anura. Frogs can secrete poison from their skin glands to prevent themselves from being eaten by predators, and they display all of the other characteristics listed. They lay their eggs in moist environments, and although they have lungs, they rely also on their moist skin as a site of gas exchange.

▌**3. (B) is correct.** The trait is sex-linked and recessive. You can tell that it is recessive because two sets of unaffected parents in the second generation produce a child that is affected. You can tell that it is sex-linked because five out of the six people affected by the trait are men.

▌**4. (C) is correct.** The movement of oxygen from the clusters of alveoli at the tips of the bronchioles in the lung, across the epithelial walls, and into the bloodstream is an example of passive diffusion. Carbon dioxide moves from the bloodstream back into the lungs, to be expelled during exhalation in the same way—through diffusion. All of the answers except *C* are examples of active transport.

▌**5. (E) is correct.** The order of tissues through which water and select minerals will pass as they travel from the roots to the xylem (then to be transported through the entire plant) is as follows: into the root hair cell, through the epidermis, through the cortex, through the endodermis (where the Casparian strip filters out unwanted substances), and then into the stele (which contains the xylem vessels).

▌**6. (A) is correct.** Bacteria are the only living organisms among the answers to evolve before algae; bacteria are prokaryotes, whereas algae are eukaryotic. The earliest eukaryotes were single-celled algae, and they are thought to have evolved about a half billion years after the prokaryotes.

▌**7. (C) is correct.** In the case of this scenario, the nerve endings in your toes acted as the receptors, receiving the information about the temperature of the water. They transmitted this information to your brain (the control center), and the brain transmitted information along a neural path to cause the effectors—the muscles working in concert—to withdraw your foot.

▌**8. (E) is correct.** Insertions and deletions are point mutations that occur when one nucleotide pair is added or lost in a gene. They have a detrimental effect on the protein product of the gene, because they affect the reading frame of the gene—all of the nucleotides downstream of the insertion or deletion will be grouped incorrectly into codons and will be misread.

▌**9. (C) is correct.** The stomata open as a result of decreased turgidity of the guard cells that flank them. As a result of this decreased turgidity, they become flaccid and the space between them becomes larger, creating an opening in the plant leaf. The change in turgidity in the guard cells is influenced by uptake and loss of potassium ions by the cells.

10. **(D) is correct.** All of the answers listed are conditions that must be met for a population to be in Hardy-Weinburg equilibrium—except the criteria that there must be only two alleles present for each characteristic. Any population that is not in Hardy-Weinburg equilibrium will evolve, and natural populations rarely achieve this type of equilibrium for extended periods of time.

11. **(B) is correct.** During mitosis and meiosis, the microtubules begin to be assembled at the centrosome, which is a cell organelle whose sole function through the cell cycle is to organize the microtubules. Animal cells have a pair of centrioles at the center of their centrosomes.

12. **(D) is correct.** The two forms that alternate in plants undergoing an alternation of generations are the sporophyte and the gametophyte. The sporophyte is the diploid form (the cells of which have two sets of chromosomes), and the gametophyte is the haploid form (the cells of which have only one set of chromosomes).

13. **(C) is correct.** Bacteria reproduce by a process called binary fission. Most bacteria contain a single, circular chromosome that also contains associated proteins. They start to replicate this circular chromosome, and one of the copies moves toward one end of the cell and at the end of replication the plasma membrane pinches inward and a new cell wall is formed, resulting in two identical daughter cells.

14. **(E) is correct.** Sources and sinks do not contribute to the movement of water and minerals up through the xylem of the plant. The movement of sugars from sources to sinks occurs in the phloem of the plant, which distributes the products of photosynthesis from the leaves down to the roots.

15. **(B) is correct.** The plant hormone responsible for promoting leaf abscission, or the loss of leaves each fall, is ethylene. This practice prevents trees from drying out in the winter when the water in the ground isn't available because it is frozen. Ethylene is also responsible for fruit ripening, and for controlling the growth of roots, leaves, and flowers.

16. **(B) is correct.** A branch from the pulmonary artery located near the lungs would be oxygen-poor. Though arteries generally carry oxygen-rich blood away from the heart, there are two exceptions, one of which is the pulmonary artery. The branches from the pulmonary artery carry oxygen-poor blood away from the heart to the lung's alveoli for oxygenation. The other exception is the pulmonary vein and its branches. They carry oxygen-rich blood away from the lung's alveoli back to the heart. The heart then pumps the oxygenated blood out of the heart to various parts of the body.

17. **(C) is correct.** Gel electrophoresis is a procedure used in the lab to separate proteins and nucleic acids on the basis of their size—essentially, the rate at which they move through a gel when an electric field is applied. The migration rate will be inversely proportional to their size, for DNA.

18. **(B) is correct.** The answer is ½. In order to deduce this, you should first figure out the gametes that each of the parents could produce. The first parent would produce the gametes *RF*, *Rf*, *rF*, and *rf*; the second parent would produce

only four *Rf* gametes. Do a Punnett square to figure out the proportions of offspring based on these gametes, and count the number of genotypes that would have the phenotype red hair and no freckles, which would be RRff and Rrff.

▌ **19. (D) is correct.** Most bacteria can be seen with a compound light microscope, whereas all of the other structures listed are too small to be seen with this type of light microscope. They must be viewed under an electron microscope.

▌ **20. (C) is correct.** Phototropism is a plant's growth in response to a light source. Negative phototropism occurs when a plant grows away from a light source, and positive phototropism occurs when a plant grows toward the light source.

▌ **21. (E) is correct.** Microtubules are involved in all of the cell functions listed except as a component of the nuclear membrane. The nuclear membrane, like all the other membranes of the cell, is composed mainly of phospholipids and associated proteins. Microtubules in the cell function in the roles of support and movement.

▌ **22. (C) is correct.** Lysosomes are digestive compartments in the cell. They are membrane-bound sacs containing hydrolytic enzymes that can digest macromolecules such as proteins, sugars, fats, and nucleic acids; the lysosomal interior has a low pH that aids in the breakdown of these large molecules.

▌ **23. (A) is correct.** Habituation is one of the simplest types of learning; it is a loss of responsiveness to a stimuli that conveys limited information or no information at all. If the dog learns that the prodding is not associated with anything—either good or bad—soon it will learn to ignore the sensation.

▌ **24. (D) is correct.** One large, visible difference between plant cells and animal cells is that plant cells have a large central vacuole. This vacuole is important to the plant because it acts as a stockroom for necessary organic compounds and is a repository for inorganic ions. The vacuole can make up about 80% of a plant's total volume.

▌ **25. (B) is correct.** One of the two general types of speciation is allopatric speciation. According to allopatric speciation, two populations are geographically separate, and gene flow has been stopped between the two populations because they are separated by distance. The other type of speciation is sympatric speciation. In this case, two populations are in the same geographic area, but biological factors (such as chromosome changes and nonrandom mating) reduce gene flow.

▌ **26. (D) is correct.** Mendel's law of independent assortment states that during gamete formation each pair of alleles will segregate independently of one another; which allele travels to which gamete is independent of the actions of the other alleles.

▌ **27. (B) is correct.** The savanna biome is characterized by vast stretches of grass with spots of trees amid them; grass is the major primary producer of this biome. The primary producer in an environment is the autotrophic organism—the one capable of turning energy from the sun into the chemical energy of food molecules.

28. (B) is correct. If you say that the brown-eyed homozygous cow has genotype *BB* and the green-eyed homozygous cow has genotype *bb*, then all of their offspring would be genotype *Bb* (blue eyed). By doing a Punnett square, you can see that crossing two individuals with genotype Bb would give you offspring in the following ratio: 1*BB*: 2*Bb*: 1*bb*. So ¼ of the offspring would have genotype *BB*, or brown eyes.

29. (E) is correct. The only adaptation listed that is not used for gas exchange is the sinuses, which are basically just spaces surrounding the organs of the body of animals that have open circulatory systems. Fish have gills, many insects have tracheal systems, many vertebrates have lungs, and some amphibians such as frogs also exchange gases across their moist epithelium.

30. (C) is correct. It is thought that birds evolved during the Mesozoic era. In common with today's reptiles, birds have amniotic eggs and scales on their feet; they differ in their feathers and adaptations for flight. The brains of birds are also significantly larger than those of reptiles.

31. (A) is correct. In reactions that are exergonic, energy is given off—often in the form of heat—during the course of the reaction. Conversely, endergonic reactions require the input of energy in order to proceed. In written reactions, the reactants are generally written to the left, and the products are written to the right of the arrow.

32. (C) is correct. During prophase of mitosis, the DNA exists in the form of chromosomes consisting of two sister chromatids. During interphase, the DNA exists as chromatin, which is loose DNA and protein together. As the cell starts to enter mitosis, the newly replicated DNA condenses into chromosomes composed of the two newly formed sister chromatids.

33. (B) is correct. O_2 receives electrons from the electron transport chain and forms water during the process of oxidative phosphorylation, and the cell makes ATP through the oxidative phosphorylation. So measuring the rate of consumption of O_2 by the cell is a good way to determine its metabolic rate.

34. (D) is correct. Transcription (the process by which DNA is transcribed into mRNA) takes place in the nucleus, whereas translation (the process by which mRNA is translated into the amino acid sequence of a polypeptide) takes place at ribosomes, which can be associated with rough endoplasmic reticulum, a series of continuous membranes in the cell. Rough ER is ER associated with ribosomes, whereas smooth ER has no associated ribosomes.

35. (C) is correct. In plant cells, both chloroplasts and mitochondria produce ATP by chemiosmosis. The thylakoid membrane of the chloroplasts and the inner mitochondrial membrane of the mitochondria and electron transport chain pumps protons across the membrane. This energy is used to power an ATP synthase that produces ATP.

36. (E) is correct. This population is exhibiting exponential growth. Exponential growth occurs whenever the conditions in an environment are ideal—when there is enough of or an excess of every resource in an environment. The population then grows at its maximum rate until it reaches its carrying capacity.

37. (C) is correct. According to the addition rule, the probability that an event will occur in two or more different ways can be calculated by adding the separate probabilities of those two ways. In this case, you can figure out the probability of alleles M and N segregating into the same gamete by adding the probabilities that either will segregate into a gamete: $\frac{1}{4} + \frac{1}{6} = \frac{5}{12}$.

38. (B) is correct. If both parents of a child are carriers of the gene that causes cystic fibrosis, their genotypes would be *Aa* and *Aa*. This means that a child produced by them would have a 25% chance of inheriting both of the recessive genes, and being *aa*. The fact that the first three children of this couple do not have cystic fibrosis would in no way affect the probability of the fourth child having cystic fibrosis—these events are unrelated.

39. (E) is correct. There has been an overall rise in the amount of atmospheric carbon dioxide, and this has been correlated with an overall rise in the temperature of the earth. Carbon dioxide traps the reflected infrared radiation from the earth in the atmosphere by absorbing it. This means that more solar heat is retained in the atmosphere.

40. (A) is correct. The groups that comprise a strand of DNA are phosphate groups, deoxyriboses, and nitrogenous bases. The four nitrogenous bases contained in DNA are adenine, thymine, guanine, and cytosine.

41. (C) is correct. Punctuated equilibrium is the term used for the idea that evolutionary change in a species occurs in rapid bursts alternating with long periods of little or no change. Gradualism is the model of evolution in which species evolve gradually and diverge more and more as time passes.

42. (D) is correct. The frog is the only animal listed that does not have an amnion at some stage in its development. An amnion is the innermost of the four extraembryonic membranes; it contains a fluid-filled sac in which the embryo is suspended. The frog doesn't need this protective sac, because it lays its eggs in an aquatic environment.

43. (B) is correct. Retroviruses are viruses that are capable of reverse transcription—they utilize an enzyme called reverse transcriptase to transcribe DNA from an RNA template. This newly made cDNA integrates into the chromosome of an animal cell and is copied along with the animal cells' DNA.

44. (B) is correct. Eukaryotic cells are generally larger than prokaryotic cells, and they are more complex. Basically prokaryotic cells have nuclear material concentrated in a nucleoid region. Unlike eukaryotes, they have no nucleus; they also lack many of the cell organelles that eukaryotes have. They are very simple cells.

45. (E) is correct. Auxins are plant hormones that are responsible for stimulating stem elongation (when they are present in low concentration), root growth, cell differentiation, and shoot branching. They also regulate the development of fruits, and they function in gravitropism and phototropism.

46. (C) is correct. Genetic engineering is the direct manipulation of genes for practical purposes, and the collective techniques under this broad field allow for the creation of hundreds of products useful to humans. Biotechnology is the manipulation of organisms or parts of organisms to make products useful to humans.

47. (B) is correct. In several important characteristics, such as nutritional mode, structural organization, growth, and reproductive technique, the fungi are more similar to animals than to plants. Molecular studies have shown also that animals and fungi are more closely related genomically.

48. (D) is correct. In humans, the anterior pituitary is responsible for the secretion of several hormones that are involved in the human reproductive cycle: oxytocin (involved in stimulating the uterus and mammary glands), follicle-stimulating hormone (which stimulates production of sperm and ova), and luteinizing hormone (which stimulates the ovaries and testes).

49. (B) is correct. RNA was thought to have been the first genetic material to evolve on earth. It is thought that RNA was self-replicating, and that the path of information flow, from DNA to RNA to protein evolved from an initial RNA system. It is thought that these RNA molecules became packaged into protobionts, which are the ancestors of living cells, and that together these evolved to become the cells of today.

50. (E) is correct. Platelets are small, enucleated blood cell fragments that are derived from bone marrow. They travel to the site of a break in the skin and release clotting factors, which through a complex set of reactions transform fibrinogen to fibrin, which in turn aggregates into threads that form a framework for a clot, sealing the break.

51. (A) is correct. The light reactions of photosynthesis convert solar energy to chemical energy in the form of ATP and NADPH. They do this when light is absorbed by various pigments in the thylakoid membrane of the chloroplasts; the pigments pass the energy down a chain of electron acceptors, and in the process ATP and NADPH are produced.

52. (D) is correct. Fungi are eukaryotes that are decomposers—also known as saprobes. They absorb nutrients from nonliving organic material like decomposing plants, dead animals, or wastes from living animals. The bodies of fungi are composed of hyphae—tiny filaments that form a mat called a mycelium.

53. (E) is correct. If a sex-linked trait is recessive, the female will express it only if she is homozygous for it, whereas a male needs only to receive the affected allele from his mother in order to be affected. If this couple produces a son with hemophilia, then either the mother or the father must carry the allele for hemophilia. The father may or may not carry the allele; however, he must transmit the y chromosome to produce a male, and the allele for hemophilia *must* come from the mother.

54. (B) is correct. This is an example of hybrid sterility—if two different species mate and produce offspring, they can still be reproductively isolated if their offspring are mostly or completely sterile. They are incapable of producing viable offspring either with each other or with either parental species.

55. (D) is correct. C_4 plants preface the Calvin cycle with a form of carbon fixation in which a four-carbon compound is produced first. The four-carbon product is exported to bundle-sheath cells, where CO_2 is again released from it, and the Calvin cycle begins to work on the CO_2. This keeps the CO_2 concentration in the bundle-sheath cells high, so rubisco accepts carbon dioxide instead of oxygen.

56. (C) is correct. The half-life of a radioactive isotope is the amount of time it takes for half of the original sample to decay. The half-life is useful because it is unaffected by temperature, pressure, or any other changes in environment.

57. (D) is correct. Pollen grains are the male gametophytes in flowering plants. In the anthers of the plant are microspores, which divide by mitosis to produce a generative cell nucleus and a pollen tube cell nucleus. Together they are enclosed in a thick wall. They create a pollen tube; the pollen tube cell nucleus and the two generative cell nuclei together are a pollen grain.

58. (D) is correct. Salivary amylase is an enzyme that is found in saliva and secreted into the oral cavity. It is capable of hydrolyzing starch (a glucose polymer found in plants) and glycogen (a glucose polymer in animal tissues). Salivary amylase breaks down these carbohydrates into small disaccharides and maltose.

59. (B) is correct. The insects are the most numerous class of animals on the earth; they live in almost all types of habitat. Their class, Insecta, contains 26 orders. Insects evolved around the Devonian period. Many insects can fly, and many undergo metamorphosis in the course of their development.

60. (A) is correct. We perceive the leaves of plants to be green because chlorophyll absorbs blue and red light while reflecting and transmitting green light.

61. (D) is correct. The fact that water molecules can form relatively strong hydrogen bonds between them results in the many unique characteristics of water. These characteristics include its being a good insulator with a high specific heat, the fact that it is more dense in liquid form than in solid form, and its high surface tension.

62. (D) is correct. In the nervous system, the nerve cells or neurons are often covered by an insulating layer called a myelin sheath. As the nerve impulse travels the length of the nerve cell, it jumps between areas not covered by myelin sheaths. These areas are called nodes of Ranvier.

63. (B) is correct. The pyruvate that is produced in the process of glycolysis (in cell respiration) is converted to acetyl-CoA, which then enters the Krebs cycle. In the Krebs cycle, acetyl-CoA is oxidized, CO_2 is reduced, and the following are produced: 1 ATP, 3 NAD+ and 1 FAD molecule.

64. (A) is correct. All of the listed animals are part of the phylum Arthropoda. Arthropods are characterized by their segmentation, hard exoskeleton, and jointed appendages. The exoskeleton of arthropods is composed of protein and chitin—and in order to grow, arthropods must shed their hard exoskeleton in a process called molting.

65. (D) is correct. Mitosis is the process by which a cell's nucleus is divided; this is followed by cytokinesis, in which the cytoplasm is divided between two daughter cells.

66. (C) is correct. Binary fission is the process by which bacteria reproduce. The bacteria replicates its DNA, and the DNA migrates to opposite ends of the cell. Then the plasma membrane infolds, and a cell wall begins to divide the two daughter cells.

67. (A) is correct. In meiosis II, the chromosome number of the cell is halved. In meiosis, the parent cell is diploid, with a chromosome number of $2n$; the daughter cells are haploid or $1n$.

68. (E) is correct. Interphase of the cell cycle is the phase during which the cell grows and replicates its genetic material. Mitosis is the part of the cell cycle during which the cell divides.

69. (E) is correct. Sharks and rays are in the class Chondrichthyes; they have cartilaginous skeletons as well as jaws and paired fins.

70. (C) is correct. Sea stars and sea cucumbers are both part of the phylum Echinodermata. These animals are slow moving and have a radial body plan. Echinoderms have a water vascular system, which is a network of canals that function in movement, feeding, and gas exchange.

71. (A) is correct. Frogs and salamanders are part of the class Amphibia. Amphibians are characterized by being both aquatic and terrestrial, as well as by laying eggs that have no hard exterior shell (which would dehydrate quickly if not laid in water).

72. (B) is correct. Reptiles such as snakes and lizards have several adaptations for land that amphibians don't have, such as scales and lungs. Reptiles are ectotherms; they absorb external heat instead of regulating their internal body temperature.

73. (D) is correct. Chordates such as humans are characterized by having a notochord, pharyngeal slits (at some point in their development), a postanal tail (again at some point in their life span), and a dorsal, hollow nerve chord.

74. (B) is correct. Chemiosmosis is an energy-coupling reaction. The energy created by a hydrogen gradient formed across a membrane is used to drive the synthesis of ATP.

75. (C) is correct. Glycolysis is the process by which glucose is split into two molecules of pyruvate. It is a metabolic pathway that occurs in all living cells and is the first part of cell respiration or fermentation.

76. (D) is correct. The Krebs cycle has eight steps and completes the breakdown of glucose started in glycolysis. In the Krebs cycle, acetyl-CoA is broken down completely into carbon dioxide. It occurs in the mitochondrion and is the second half of cell respiration.

77. (A) is correct. An electron transport chain is composed of a series of electron carriers (which are proteins) embedded in a membrane. They shuttle electrons and in the process release energy that is used to make ATP.

78. (E) is correct. In the light reactions of photosynthesis, solar energy is converted to the chemical energy of ATP. The light reactions of photosynthesis take place in the thylakoid membrane in chloroplasts.

79. (C) is correct. A species is defined as a population or group of populations whose members can interbreed with one another to produce viable, fertile offspring.

80. (D) is correct. The niche an organism occupies in its environment is defined as the sum of the biotic and abiotic factors and resources a certain species utilizes.

81. (A) is correct. A population is defined as a group of individuals of one species that live together in a certain geographic area.

82. (B) is correct. A community consists of all of the organisms that live in a particular geographic area. It consists of all of the populations in the area.

83. (E) is correct. Macrophages are what monocytes develop into—monocytes are leukocytes or white blood cells. Macrophages are the largest phagocytotic cells, and they are long-lived.

84. (D) is correct. Eosinophiles are a type of leukocyte that act against large parasitic invaders; they position themselves against the wall of a parasite and inject destructive enzymes into the invader.

85. (B) is correct. Antibodies are secreted by B cells, and they are proteins that are specific to antigens. They represent the effectors in an immune response.

86. (A) is correct. Antigens are foreign particles in the body that elicit a response from the immune system.

87. (C) is correct. Histamines are chemical signals that are released by cells of the body in response to an injury; they are produced by basophils and mast cells.

88. (B) is correct. The stomach is an elastic muscular sac that stores and digests food. It secretes gastric juice, which starts the initial digestion of proteins along with the churning motion of the stomach walls.

89. (D) is correct. The gall bladder is an organ found in the liver that stores bile and releases it when needed into the small intestine to emulsify fats.

90. (E) is correct. The duodenum is the first section of the small intestine; here acid chyme from the stomach mixes together with digestive juices from the pancreas, liver, and gallbladder.

91. (C) is correct. The pancreas secretes digestive enzymes—as well as an alkaline solution—into the small intestine.

92. (A) is correct. The liver produces bile, and bile is stored in the gall bladder. It acts as a detergent, aiding in the digestion of fats.

93. (B) is correct. The phospholipid bilayer is made up of phospholipids, which have a hydrophilic head group and two hydrophobic, fatty acid chains. The head group points outward to the cytoplasm.

94. (C) is correct. The hydrophobic fatty acid tails of the phospholipids point inward, avoiding contact with the cytoplasm.

95. (A) is correct. Proteins are both embedded in the phospholipid bilayer (these are integral proteins) and associated with the cytosol face of the cell membrane (these are called peripheral proteins).

96. (C) is correct. The enzymes that would be able to function best in the bloodstream of humans according to this data are 1, 2, and 5. The temperature of the human body is about 36–38 degrees Celsius, so enzymes 1 and 2 would be somewhat functional; the pH of the bloodstream is about 7.4, so only enzyme 5—and not 4—would function at that pH.

97. (E) is correct. The enzyme that has the optimal activity at pH 8 would work best in the small intestine, and this enzyme could be trypsin. The small intestine is slightly alkaline (or basic) in pH, so the enzyme you look for on the graph should have its optimal activity in the alkaline range.

98. (C) is correct. The enzyme most likely to function in the hot springs of Yellowstone Park is enzyme 3, which has an optimal activity at a temperature of about 80 degrees Celsius. Enzymes of thermophilic (heat-loving) bacteria work best at very high temperatures.

99. (E) is correct. Enzyme 5 is the most efficient enzyme here—it has the highest rate of reaction. This means that its optimal rate is faster than that of any of the other enzymes depicted, assuming a standard measurement for the y axis of both graphs.

100. (A) is correct. Just after the ^{238}U was injected into the left atria, it would travel to the left ventricle. Blood enters the heart through the right and left atria; next it travels to the right or left ventricle.

101. (B) is correct. Just after it was injected, the blood carrying the radioactive carbon would be detected dropping off oxygen in the systemic capillaries. Blood leaves the left ventricle, enters the aorta, and then travels through the body, dropping off oxygen to active metabolic tissues.

102. (A) is correct. Shortly after it was injected, the ^{3}H isotope would be found in the right ventricle, since blood is pumped from the atria to the ventricles first.

103. (C) is correct. Just after it was injected, the blood carrying ^{32}P would be found in the lung capillaries, where it would be picking up oxygen and dropping off carbon dioxide before returning to the heart through the left atria. Oxygen-depleted blood is pumped through the right side of the heart on its way to the lungs.

104. (E) is correct. There is a peak in the rate of photosynthesis at around 680–700 nm of light wavelength because photosystem I contains reaction-center chlorophyll that best absorbs light at around 700 nm, whereas photosystem II contains reaction-center chlorophyll that best absorbs light at around 680 nm. The absorption of light is the first step in photosynthesis.

105. (B) is correct. An absorption spectrum is a graph plotting a particular pigment's light absorption—that is, the fraction of light not reflected or transmitted versus the wavelength of light. This is a plot of the absorbency of the chlorophyll *a* pigment vs. light wavelength.

106. (D) is correct. The reason why the action spectrum for photosynthesis doesn't match the absorption spectrum for chlorophyll *a* is because chlorophyll *a* is not the only photosynthetically important pigment in the chloroplast. Two other photosynthetically important pigments are chlorophyll *b*, and carotenoids.

107. (B) is correct. Species A is most likely to be autotrophic—autotrophs are capable of converting energy from the sun into chemical energy of organic molecules. As you can see, A does not consume any other species (likewise with B), so it must produce its own food by obtaining energy from the sun.

108. (D) is correct. Species E, F, and G are all secondary consumers in this food web. Secondary consumers consume primary consumers, and primary consumers consume primary producers. In this case, for example, species F eats species C and D, which in turn consume primary producers A and B.

109. (B) is correct. The carnivores in this food web are represented by species F, G, H, and I. Carnivores are generally secondary, tertiary, and quaternary consumers, and they are distinct from omnivores, which eat both plants and animals. Species E is an example of an omnivore.

110. (E) is correct. If this were a food web drawn to represent the ecosystem of a savanna, species A and B most likely would represent two species of grasses, since grasses are the predominant primary producers of the savanna biome. Insects are also predominant as primary consumers, so species C, D, and E could represent different species of insects.

111. (B) is correct. Species B probably has the most guanine-cytosine base pairs of any of the other species' DNA, because it has the highest heat of denaturation. Since there are three hydrogen bonds that form between cytosine and guanine, and only two between adenine and thymine, DNA strands with more guanine-cytosine base pairs would be harder to separate. Thus more energy in the form of heat would have to be put into the system.

112. (D) is correct. The two species that are likely to be most closely related evolutionarily are D and E. Since their DNA denatures at about the same temperature (57°C and 58°C, respectively), this means that the composition of their DNA might be similar.

113. (C) is correct. Another way to denature DNA is to lower the pH of their environment—DNA denatures at low pH, and keeping track of the pH at which each of the DNA samples degraded would give you the same kind of data as the table above. DNA that degraded at relatively higher pH would be less tightly bound than DNA that degraded at lower pH.

114. (E) is correct. E represents the total energy of the transition state of the uncatalyzed reaction. The transition state is the stage of the reaction where enough energy has been absorbed by A and B to surmount the activation energy; this state represents the least stable point in the reaction.

115. (A) is correct. This reaction is exergonic—that is, the energy of the products is lower than the energy of the reactants. This means that during the course of the reaction, energy is given off to the surroundings, perhaps in the form of heat.

116. (B) is correct. B represents the activation energy of the enzyme-catalyzed reaction. The activation energy is the initial investment of energy needed to start a reaction. This activation energy is usually provided in the form of heat, which is used to break the bonds of the reactants so that they can re-form in the correct configuration to form the product molecule(s). As would be expected, the activation energy of the catalyzed reaction is lower than that of the uncatalyzed reaction (C).

117. (D) is correct. The difference in energy between the reactants and the products of this reaction would remain the same, whether this reaction were catalyzed or not. Catalysts speed up reactions by decreasing the activation energy of the reaction, but they do not affect the amount of free energy contained in either the reactants or the products.

118. (B) is correct. The best advice for this farmer would probably be to replace his crop of C_3 plants with a crop of C_4 plants. C_4 plants use a type of photosynthesis that is more efficient in hot, arid climates, and if the climate of his farm has been getting steadily more hot and arid, then he should plant a crop that thrives under those conditions.

■ **119. (D) is correct.** Nitrogen is an important macronutrient in plants. It cannot be absorbed from the atmosphere by plants, so it must exist in the soil and be taken up by plant roots. Often plants have nitrogen-fixing bacteria associated with their roots. These symbionts obtain nutrients from the plant roots while converting nitrogen into a form that is useable by the plant.

■ **120. (C) is correct.** Most of a plant's mass is derived from the carbon dioxide it takes in from the atmosphere. Through the process of photosynthesis, it converts this carbon dioxide into sugar molecules, which are transported throughout the plant and which comprise most of its mass.

Free-Response Questions

1. (a) The cell membrane of a typical animal cell is composed of three main things: phospholipids, which are two fatty acids connected to two glycerol hydroxyl groups and a phosphate group connected to the third glycerol hydroxyl group; proteins, both integral (embedded in the cell membrane) and peripheral (associated with the outside of the membrane); and membrane carbohydrates and glycolipids, which are small carbohydrates associated with the outsides of the membrane. One other component of the cell membrane is integral carbohydrates, which act along with unsaturated fatty acid chains of some of the phospholipids to make the membrane more fluid and less rigid.

The function of the phospholipids is to create a loose background foundation for the rest of the molecules in the cell membrane. Some very small molecules and ions can pass through the lipid membrane unaided. This type of movement across the membrane is called passive diffusion, because the cell doesn't expend any energy getting the substances across.

The function of the proteins is multi-fold, but one important function they have is to facilitate the passive transport of water and certain other solutes across the membrane. The proteins that serve this function are called transport proteins. Proteins can also participate in the active transport of certain substances across the membrane; they can act as pumps that use the power of ATP to transport substances against their concentration gradient. Proteins can also act as important cell-surface receptors.

Membrane carbohydrates and glycolipids on the cytosolic surface of the cell membrane are important in cell-cell recognition; these carbohydrates and glycolipids vary from species to species, and from cell type to cell type, so they enable cells to tell each other apart.

(b) The way viruses get their DNA into an animal cell varies from virus to virus—all have different techniques. One type of virus, the T phages, uses its tail apparatus to inject DNA straight into the bacterium's cell membrane. Viruses identify the specific type of cell they need to infect through the protein receptors on the cell's surface.

Hormones are the chemical messengers of the body. They travel through the bloodstream until they reach their target cell. There are two ways by which hormones can gain entry into a cell. The first method is for steroid-based hormones. The hormone diffuses first through the cell membrane, then through the cytoplasm, and then it binds to a specific receptor protein in the nucleus. The second method is used for peptide-based hormones. The hormone binds to a specific receptor protein found on the target cell membrane. This binding signals a series of biochemical signal-transduction pathways.

Water enters the cell through a process called facilitated diffusion. This means that water crosses the cell membrane down its concentration gradient, but with the help of specific transport proteins. Transport proteins are specific for the molecules they help across the membrane, but they do not require the input of energy.

This is a good free-response answer because it shows knowledge of the following important biological terms:

phospholipid	*facilitated diffusion*
fatty acid	*active transport*
glycerol	*T cell*
integral protein	*hormones*
peripheral protein	*target cell*
carbohydrate	*signal-transduction pathway*
glycolipid	*receptor*
passive diffusion	

The response also shows knowledge of the following important biological processes—the importance and function of the cell membrane; how molecules get across the cell membrane, how viruses get their DNA into a cell; and how hormones affect target cells.

2. (a) When plants moved onto the land, one major obstacle they faced was dehydration. Plants that are submerged in water do not need any adaptations for conserving water (because water is all around them and available at all times), but land plants are in danger of losing water through evaporation. Plants faced a reproductive obstacle—fertilization—as well. In water, the male and female gametes could float to another plant and land on it, and then fertilization could occur. On land, however, how could the gametes reach each other—and then how could the zygote be protected from dehydration? A third obstacle to plants' living on land was structural. In aqueous environments, plants are supported by the water around them. On land, they would need to support themselves somehow, but they didn't yet have tough, supportive tissues.

(b) Plants solved the problem of conserving water through the evolution of an epidermal layer on the outside of their leaves that is covered with a cuticle—that is, a layer of waxy polymers. The cuticle serves to protect the plant and prevents evaporation of water through the epidermis. Plants also have stomata, which allow for exchange of oxygen and carbon dioxide and are needed for photosynthesis. Stomata are capable of opening and closing to control excessive transpiration. Terrestrial plants also evolved vascular tissues for transporting water and other nutrients around the plant body, and these vascular tissues are xylem and phloem. Xylem will transport water from the roots up to the leaves, and phloem will carry sugar from the leaves to various parts of the plant.

The problem of how to reproduce on land was solved in part by the evolution of the seed. The seed is the plant embryo combined with a food supply encased in a protective coat. This protective coat prevents the embryo from dehydrating even if the seed sits on the dry ground for a relatively long period of time. Another adaptation to aid in plant reproduction on land was the spore. Spores are reproductive cells that can develop into a mature plant without fusing with another cell. They are generally lightweight and can travel significantly far from the parent through the air and grow into a new plant.

Plants adapted to living on land without the structural support of water. They developed hard, stiff shoots that enabled them to grow to great heights. The shoots of plants are made up of several types of plant tissues, including the vascular tissues xylem and phloem. Both xylem and phloem are made up of dead water-conducting cells joined together to form long, stiff tubes that help support the plant in its growth.

This is a good free-response answer because it shows a working knowledge of the following important terms:

fertilization	*seed*
gametes	*xylem*
epidermis	*phloem*
cuticle	*spores*
stomata	

The response also shows a working knowledge of the important biological concept of the evolutionary adaptations of plants that enabled them to colonize terrestrial environments.

3. The cell cycle of a plant cell, much like the cell cycle of most other types of cells, includes two main phases—interphase and mitosis. Interphase is divided into three phases: the G_1 (gap 1), S (synthesis), and G_2 (gap 2). The cell undergoes a tremendous amount of biochemical activity during the G_1 phase, where the plant cell grows and produces new organelles. In the S phase, synthesis of new DNA material takes place. During the G_2 phase there is continued growth, and organelles needed for cell division or mitosis are replicated. There are several checkpoints in the cell cycle of a plant cell, the most crucial of which is the checkpoint. If the plant cell gets the go-ahead signal at this checkpoint, it will be committed to divide. If not, it will enter the G_0 phase, which is a nondividing phase.

In late interphase, just before the mitotic phase, the nucleus is intact and the chromosomes are not well defined in the nucleus. A pair of centrioles has been made in the cytosol from the replication of one centriole, and it awaits its role in cell division. When the plant enters the first mitotic phase (prophase), the chromatin fibers become more condensed and begin to look like chromosomes, each of which has two sister chromatids. In the cytoplasm, the mitotic spindle forms, and the centrioles move away from each other and toward the opposite poles of the cell. The next phase that the plant cell would enter is prometaphase, in which the nuclear envelope would fragment and the microtubules would attach to the kinetochores of the chromatids. In metaphase, all of the chromosomes would line up on the metaphase plate at the equator of the cell, and microtubules would be attached to the kinetochores of every sister chromatid. When the plant cell entered anaphase, the sister chromatids are "dragged" apart from each other to opposite ends of the cell by the retracting microtubules.

The final stage of plant cell mitosis is telophase and cytokinesis. In telophase, two new nuclei begin to form around the groups of sister chromatids that have now migrated to opposite ends of the cell. The chromosomes becomes less condensed, and the cell plate—which divides the cytoplasm in two—starts to grow at the center of the cell, eventually growing all the way to the perimeter of the parent cell and effectively dividing the cell in two.

Following the outline above, this response deals with the following important terms:

cell cycle	chromatin fibers
G_1 phase	sister chromatids
S phase	mitotic spindle
G_2 phase	prometaphase
G_1 checkpoint	metaphase
G_0 phase	nuclear envelope
nondividing phase	microtubules
interphase	kinetochores
mitotic phase	telophase
centrioles	cytokinesis
prophase	cell plate

The response also describes thoroughly the following processes—the cell cycle including its checkpoints, and plant cell mitosis and cytokinesis.

4. (a) There are three convincing reasons why the theory that glycolysis was the first ATP-producing metabolic pathway to evolve is probably true. The first reason is that long ago Earth's atmosphere contained almost no oxygen, and only relatively recently have the current atmospheric levels of gases come to be what they are. Glycolysis does not require oxygen, so it is possible that prokaryotes (which evolved before eukaryotes) utilized this method for making ATP.